NICHE HIERARCHY:
Structure, Organization, and Assembly in Natural Systems

George Sugihara

J.ROSS
PUBLISHING

ISBN-13: 978-1-60427-128-7

Printed and bound in the U.S.A. Printed on acid-free paper.

10 9 8 7 6 5 4 3 2 1

Library of Congress Cataloging-in-Publication Data

Sugihara, G.
Niche hierarchy : structure, organization, and assembly in natural systems / by George Sugihara.
pages cm
Originally presented as author's Ph.D. dissertation, Princeton University, 1983.
Includes bibliographical references.
ISBN 978-1-60427-128-7 (hardcover : alk. paper) 1. Niche (Ecology) 2.
Hierarchies. 3. Birds--Behavior. 4. Biotic communities. 5. Habitat (Ecology) I. Title.
QH546.3.S84 2015
577.8'2—dc23
2015030036

Direct all inquiries to J. Ross Publishing, Inc., 300 S. Pine Island Rd., Suite 305, Plantation, FL 33324.

Phone: (954) 727-9333
Fax: (561) 892-0700
Web: www.jrosspub.com

CONTENTS

FOREWORD

by Lord May

I had the privilege and pleasure of being George Sugihara's supervisor for his Ph.D. at Princeton University. This book is indeed his thesis, and I am delighted that it is now more easily available.

George was a notably independent and creative student. His thesis research—and thus this volume—began by seeking to understand the general patterns that are observed in the relative abundance of species. This in turn led to an examination of the hierarchical organization of species' niches, and the way in which such considerations can explain much of what we observe in distributions of the relative abundance of species.

In short, this book is organized as a series of essays, with the aim essentially being to illuminate the interrelations among the properties of "topological structure, dynamics, and relative abundance in multi-species communities."

I particularly like Sugihara's "new mathematical characterization of the niche," which is roughly aligned with Hutchinson's (1958) concept of a hypervolume, but with the difference that here a species is thought of as an n-dimensional set of resources, rather than in the more conventional way as an "n-dimensional hypervolume *per se*."

This leads on to indicate how such an "architectural" definition of a species' niche poses the question of whether or not there are holes in niche space. There is a clear connection here with the much-debated "species packing problem," posing the question of whether there may be minimal constraints on niche structure. Sugihara confronts these ideas with "coarse data from a diversity of food webs, and more refined data from bird communities." His interesting finding is that most, if not all, real such assemblies tend to be packed in a conservative manner, not allowing any "holes."

In the fifth of the book's seven chapters, Sugihara focuses on the question of whether niche space is organized hierarchically; that is "whether species niches are strung out together in a tree-like fashion." The answer to the question, thus posed, is found by Sugihara to be that the real niche spaces thus examined are indeed consistent with organization of the above kind.

The concluding, seventh chapter, focuses on the relationship between hierarchical community structure and species patterns of abundance. The aim here is to see to what extent there is a connection between two distinct ideas about community structure: namely, functional organization and the abundance distribution of species. In a way that I find most engaging, Sugihara's examination of this question shows that hierarchical organization in the niche can account for most of the abundance structures that are seen in practice. In particular, these considerations explain Preston's (1962) canonical lognormal distribution, Motomura's (1932) geometric series, and MacArthur's (1957) broken-stick distribution. The book also presents interesting data from bird communities, which lead to similar tests of the correspondence between functional organization and relative species abundance in ecological communities.

Professor Lord May of Oxford, OM AC Kt FRS
Department of Zoology, University of Oxford

FOREWORD

by Jordi Bascompte, Ph.D.

It is a pleasure to have the chance to add a few words to this book. I have three reasons for being thrilled about this task. First, although the context of this text goes back to 1982, it has remained largely hidden as it was never published. Thus, I can contribute to uncovering a hidden jewel. Second, the content of this text remains as valid and fresh today as it was more than 30 years ago, which is a distinction of extraordinary work. Third, I was myself very much influenced by George Sugihara's vision about the structure of ecological communities long ago when I was starting my own Ph.D. Is there anything more fulfilling that being able to attach one's name to the work that inspired him? This text, therefore, serves to pay my gratitude for the inspiring readings. Indeed, science is a process of continuous influence.

I am sure the broad scientific community will appreciate having the chance to learn about these ideas. Indeed, brilliant ideas never get old; on the contrary, they just become more and more relevant as one can see them helping unfold a research avenue. Thus, one interesting opportunity brought by the publication of this book is to rethink how these ideas have contributed to the development of community ecology. I encourage the reader to go through these pages and to try to relate several of the ideas here described with relevant directions in today's ecology. It will not be a difficult exercise, and at the end, the ultimate proof that good thinking is never out of date.

One theme that transcends across Sugihara's work is the belief that there are basic organizational principles underlying natural communities despite the obvious variation in species composition, species richness, and life-history strategies. This view has the power to uncover basic and yet powerful mechanisms explaining the patterns we see in natural communities. This way of thinking is as valid today as it was in 1982. It is an intellectual approach that goes beyond the accumulation of facts since it requires a creative mind. This is similar to the type of theoretical work done in physics. As with the good work in physics, one has the feeling that Sugihara's work is eminently beautiful. And this was the most vivid feeling I recall when being exposed for the first time to this work at the early stage of my career. Indeed, beauty has been foreseen as the hallmark of truth in theoretical physics.

More specifically, I would highlight three main ideas advanced by Sugihara that—irrespective of the specific format they are presented—continue to be of extraordinary importance. First is the idea that community structure affects its dynamics. This is at the root of all research on ecological networks as its rationale is very much based on the idea that understanding the architecture of the web of life is important to understand its robustness to perturbation—an extremely timely topic in the face of global change. It is increasingly clear that there are constraints in the architecture of natural communities that can make them more robust. What is not so clear, though, is whether the need for populations to be both feasible and persistent is the driving force beyond such topological patterns. This could certainly be the case, but it may also be that the observed network patterns are just a byproduct of different mechanisms that may secondarily enhance community persistence. This latter scenario would be analogous to how the concept of "exaptation" expanded the adaptationism program in evolutionary biology.

Second, Sugihara's work makes very clear the idea that there are simple mechanisms of community assembly that can be tested with current patterns of species abundances. This is relevant when understanding how communities were built. It conveys the idea that the complexity we observe in natural communities may have simple generative rules. This is very much aligned with current work in theoretical ecology that echoes similar developments in the complexity sciences.

Third, this book is very much related to a geometric description of species coexistence. It allows looking at one of the classic questions in ecology—why are there so many species?—from a novel perspective. One key concept in this geometric description is the notion of hierarchy, a powerful organizational scheme that relates different species across simple and meaningful axes. This concept, not surprisingly, is also relevant in other fields of science. For example, hierarchy and invariance across scales was a key topic during the development of fractal theory.

In summary, this work reads as exciting now as it read back in 1982. I expect the book will spur the imagination and fascination of many young scientists, while helping more senior ones to carve a deeper conceptual framework to place current concepts and results.

Happy reading!
Jordi Bascompte, Ph.D.

PREFACE

The main theme is stated in the beginning: *"If deep general principles are to be found (in ecology) they are likely to be associated with general widely observed empirical regularities.... and that the theory should be as directly and simply connected to measurements as possible."* This opinion still sounds true to me, and the question of how to use available data to render one of the cornerstone ideas in ecology— Hutchinson's n-dimensional niche—has remained a central vision.

What is presented here is a heuristic data-driven theory at a high level of abstraction that resonates more with the intellectual traditions of physics than those commonly exercised in ecology. The style of thought is physics-like and agnostic in that it tries to put data parsimonious theory before current preconceptions. It is minimalist and takes as a starting point an attempt to unify two disparate areas in ecology for which data were most available: the classic study of species abundances and the study of food webs. It uses these data with minimal assumptions to reveal patterns about the structure of niche space that suggest unifying high-level mechanisms.

Of note is that fact that the work makes several predictions that were unknown at the time but have subsequently been validated, such as a negative skew in a fully unveiled lognormal species abundance distribution (Nee et al., 1993), and a connection between diagrams of niche similarity and species relative abundance (Sugihara et al., 2003). Both predictions are a product of hierarchically structured niche space. Also, the attempt in Chapters 4 and 6 to connect static topological niche patterns to the process of community assembly and to system dynamics and stability, resonates with continuing works on these relationships and helps to set the stage for an entirely different approach to understanding the communal niche—work based on rendering time series data as an attractor in an n-dimensional state space—a dynamic version of Hutchinson's n-dimensional niche.

George Sugihara

ABSTRACT

This collection of essays is an effort to link three major areas in the theory of large multispecies communities: topological structure of species interrelationships, population dynamics and species relative abundance.

Chapter 2 begins by introducing a new mathematical characterization of the niche that aligns roughly with Hutchinson's hypervolume concept but where a species is viewed as an n-dimensional resource set (resource clique or simplex) rather than an n-dimensional hypervolume. In geometrical terms these species niches may be thought of as convex polyhedra whose vertices correspond to resources. This generates a powerful architectural description of communities where the polyhedral species niches are joined through shared resource vertices.

Chapter 3 applies this characterization to determine whether natural ensembles are packed densely as a simple solid or whether they are constructed loosely to contain holes as in a multidimensional swiss cheese. This is viewed as the converse of the well-known species packing problem and asks if any minimal constraints come into play in assembling communities. An analysis of data from 73 real food webs and more refined data from 11 bird communities suggest that most if not all real communities are packed in a conservative and dense manner that allows no holes. This is a fundamental result about the topology of niche space.

Chapter 4 centers on Cohen's provocative finding that nature contains an excess of interval food webs. This excess is shown to be a by-product of a more fundamental property in the structure of niches, the rigid circuit property, from which it is possible to deduce a simple and biologically sensible rule operating in the assembly of communities. This mathematically deduced assembly rule is shown to be capable of explaining the observed high frequency of interval and rigid circuit graphs as well as the absence of holes in the niche discovered in the previous chapter.

Chapter 5 asks whether niche space is organized as a structural hierarchy, i.e., whether species niches are strung together in a tree-like fashion. By relating this formally to the question of holes, most of the real niche spaces examined are found to be consistent with organization of this kind. Also introduced here is the notion of functional hierarchy where the binary linkages of the structural portrait are given weight and degree, so that communities can be visualized as a dendrogram composed of nested guilds according to increasing degree of functional similarity. Results demonstrating a correspondence between functional hierarchy and properties of the underlying structural portrait are discussed by introducing the concept of compatible ordering.

Chapter 6 begins to forge the link between topology and dynamics and investigates the cause-and-effect relationships between observed topological patterns and their dynamic background. Preliminary results here show that observed structural patterns in niche overlaps may indeed be at least partially shaped by dynamic behavior, and that such organization in ecological communities in turn helps to place them in the sparse regions of parameter space possessing enhanced stability properties.

Finally, Chapter 7 examines the consequences of hierarchical community structure for species abundance patterns. The motive here is to seek a correspondence between two separate classical ideas about structure: functional organization and species abundance distributions. Here it is shown that hierarchical organization in the niche may account for most observed abundance structures including Preston's (1962) canonical log-normal distribution, Motomura's (1932) geometric series and MacArthur's (1957) broken-stick distribution. Data from bird communities are used to make independent and more restrictive tests of the correspondence between functional organization and relative species abundance.

ACKNOWLEDGMENTS

I am deeply indebted to many people for their patience and guidance.

Above all I would like to express profound gratitude to my advisor, Dr. Robert M. May, whose counsel, encouragement and mastery of the "gentle nudge" were indispensable in helping to see the project through. It is always an honor to be able to work with one's hero in the field, and to discover that behind the brilliance is a truly wise and compassionate man.

Dr. Ruth Patrick also played a central role in stimulating the project. It was largely her early work on lognormal distributions that inspired my interest in species abundance patterns, and I am grateful for the many engaging opportunities to learn at the knee of one of the great ecologists of our time. Many of the ideas in Chapters 5 and 7 I owe directly to our numerous conversations at the Academy of Natural Sciences in Philadelphia.

Other teachers and colleagues whom I would like to thank include: Dr. Daniel A. Livingstone, Dr. Robert Whittaker, Dr. Henry Horn (for his excellent suggestions on revisions), Dr. Harold Hastings, Dr. Frederick Briand (for kindly providing the data used), Dr. Andrew Siegel, Dr. Peter Honeyman, Dr. John Lawton, Dr. Norman Mac Donald, and Dr. Steven Schulman. Special thanks to Dr. Andrew Siegel for our Wednesday luncheons and warm intellectual companionship, and to Dr. Peter Honeyman of Bell Laboratories for many days of sheer joy (and good soup) working on the computing problems in the thesis. Without Dr. Honeyman's friendship and collaboration, many of the numerical results in Chapters 3, 4, and 6 would not have been possible.

Sincere thanks also to Joan Hall for the beautifully drawn figures in the text and for assistance in rewording some of the more complex ideas in simple terms. Pierre du Pont helped greatly in creating the expert computer graphics in Appendix B.

1

THE BEGINNING

If deep general principles are to be found (in ecology) they are likely to be associated with general widely observed empirical regularities. This work is motivated by the idea that conceptions and theory should arise from observations and measurements of nature, and that the theory should be as directly and simply connected to measurements as possible.

1.1 INTRODUCTION

Three important properties of a large multispecies community are the topological structure of species interrelationships (e.g., who competes with whom), the dynamic nature of these interactions, and the resulting pattern of species relative abundance. Ideally, a logical order in which these properties might be studied would be first to obtain a sketch of the bare outlines of community organization by determining the topological constraints acting within real communities, then to color in the sketch by adding dynamic behavior to the static linkages among species, and finally to extend these insights toward understanding macroscopic patterns of species commonness and rarity. Each step, by this plan, leads naturally to the next in terms of detail and complexity.

In contrast to this ideal situation, the historical arrow has actually flown in the opposite direction. Species abundance patterns which figured prominently since the early part of this century were superseded in the 70's by concern over the dynamic behavior of ecological systems, and it is only recently that attention has focused on a formal study of the topological properties of communities. This realized trajectory is understandable and corresponds both to the conceptual primacy of these properties as well as to the chronological order in which the mathematical tools required for each subject were refined. Dominance and diversity are certainly more directly observable than are fluctuations in population sizes, and these in turn are less subtle to perceive than the linkage patterns of niche interrelationships. The tools for probing structure, graph theory and algebraic topology, have really only come into their own in the last few decades, compared with the better established disciplines of differential equations and statistics used to study dynamics and relative abundance patterns.

There is little doubt that the reversed ordering of events was necessary, however an apparent consequence of the historical trajectory is that connections between these three major areas are either weak or completely missing. As Robert May (1979) recently remarked, the relationships between community organization and dynamics is largely speculative and have yet to take advantage of real topological constraints such as those discovered by Cohen (1978) in the empirical data. The implications of structure and dynamics for species abundance patterns is presently also a wide open area for inquiry. This seems a bit curious especially in view of the effort devoted by earlier generations of ecologists to the species abundance problem. Whether these discontinuities are an expression of specialization, disenchantment with unyielding problems, or popular trends, it seems likely, had the flow gone from topology to abundance patterns, or from dynamics to topology that a much more unified picture would have emerged by now.

1.2 MENU

This collection of essays attempts to tie together the properties of topological structure, dynamics, and relative abundance in multispecies communities. It is anchored at either end by two provocative empirical regularities: Cohen's (1978) discovery that real ecological ensembles can be rendered in one-dimension, and thereby represent a small subset of mathematically possible topologies, and Preston's (1962) finding that most real taxonomic communities can be uniquely described by a specific one-parameter relative abundance distribution, the so called canonical lognormal distribution of species abundance. In addition, a new empirical regularity is introduced showing that niche space has the topology of a simple solid rather than of a perforated multidimensional swiss cheese. The idea used to unify these properties is the broad concept of hierarchical organization which is developed in two senses: structural hierarchy and functional hierarchy. Structural hierarchy refers to the formal topology of trophic linkages and functional hierarchy to the dendrogram of relationships organizing the system.

Chapter 2 begins by presenting a new mathematical characterization of the niche which aligns roughly with Hutchinson's (1958) hypervolume concept but where a species is viewed as an n-dimensional resource set (resource clique or simplex) rather than an n-dimensional hypervolume *per se*. This makes Hutchinson's idea operational with available foodweb data and allows for a sensible discussion of the communal niche space—sensu Whittaker's (1972) community hypervolume—in cases when the species involved have resource sets of different dimensionality.

Chapter 3 shows how this architectural description of the niche can be used to make sensible the question "are there holes in niche space?" This is presented as the converse of the well-known species packing problem and asks whether there are any realized *minimal* constraints on the structure of the niche. Tested on coarse data from a diversity of food webs, and more refined data from bird communities we find that most if not all real ensembles tend to be packed in a conservative manner that allows no holes.

Chapter 4 centers on Cohen's finding that nature contains an excess of interval food webs. This excess is shown to be a by-product of a more fundamental property in the structure of niches, the rigid circuit property; which, in turn, leads to a simple and biologically sensible rule operating in the assembly of communities. The assembly rule is derived by mathematical deduction and is necessary and sufficient to explain the observed high frequency of rigid circuit graphs as well as the absence of holes in the niche discovered in the previous chapter.

Chapter 5 asks whether niche space is organized as a structural hierarchy i.e. whether species niches are strung together noncircuitously in a tree-like fashion. By relating this formally to the question of holes, most of the real niche spaces examined are found to be consistent with organization of this kind. Also introduced here is the notion of functional hierarchy where the binary linkages of the structural portrait are given weight and degree, and the ensemble is assembled into nested guilds, chinese box fashion, according to increasing degree of functional similarity. Results demonstrating a correspondence between functional hierarchy and properties of the underlying structural portrait are discussed by introducing the concept of compatible ordering.

Chapter 6 begins to forge the link between topology and dynamics and investigates the cause and effect relationships between observed topological patterns and their dynamic background. Preliminary results here show that observed structural patterns in niche overlaps may indeed be at least partially shaped by dynamic behavior, and that such organization in ecological communities may in turn help to make them more robust to outside perturbations.

Finally, Chapter 7 examines the consequences of hierarchical community structure for species abundance patterns. The motive here is to seek a correspondence between two separate classical ideas about structure: functional organization and species abundance distributions. Here it is shown that hierarchical organization in the niche may account for most observed abundance structures including Preston's (1962) canonical lognormal distribution, Motomura's (1932) geometric series and MacArthur's (1957) broken stick distribution. Data from bird communities are used to make independent and more restrictive tests of the correspondence between functional organization and relative species abundance.

This work is not intended to be a comprehensive treatment of topology, dynamics and the species abundance problem. In particular, one specific aspect of niche topology not discussed is the directed-graph or food-web portrait of predator prey relationships investigated recently by Pimm and Lawton (1980), Auerbach (1982), and Post and Pimm (1982). Rather than focusing on the predator prey picture we will concentrate on the competitive structure of niche overlaps, mainly because this is a level where interesting regularities appear and find coherent expression.

Although the treatment of dynamics uses simple analysis based on stability and equilibrium, the results reported here are believed to be robust to these assumptions. The link between dynamics and species abundances remains indirect in these essays and is mediated by the extent to which dynamics helps to explain observed topologies. A further study that connects explicit ecosystem dynamics to abundance patterns and to fluctuations in abundance remains a focus of future work.

2

CHARACTERIZING THE NICHE

Our aim here is to develop a mathematical characterization of the ecological niche that is precise and actionable and that can be rendered with existing food web data. A fundamental requirement for this exercise is having a judiciously chosen set of *objects* and *relations*. As we shall see, the view into the structure and organization of ecological communities will depend importantly on the particular objects and relations chosen to describe the niche.

2.1 A PROBLEM OF INCOMPATIBLE DIMENSIONS

"Ecological niche" like many other fundamental terms in ecology has no simple agreed upon definition (see Hurlbert, 1980, for a recent review). One idea that rises above the variety of definitions that have been proposed is Hutchinson's (1958) geometrical n-dimensional hypervolume. According to this definition a species niche is viewed as a region in a hyperspace defined on axes representing all of the ecological factors determining its existence. As a conceptual framework this is an idea that has figured prominently in virtually all subsequent uses of the term (e.g., see Colwell and Fuentes, 1975), and stands firmly placed as one of the cornerstones of modern ecological theory.

An extension of Hutchinson's geometrical view from the level of individual species to the whole community was attempted by Whittaker (1972), who introduced the notion of a community hypervolume. Here the coordinate axes represent the totality of ecological requirements for the *entire* community, rather than for just a single species. In broadening the compass of the Hutchinsonian definition to the community level, Whittaker created a powerful metaphor for discussing the evolution of community structure in terms of how species niches partition and separate within the hypervolume over time.

The community hypervolume is a compelling metaphor, however, as an analytical tool there are operational difficulties. One particular problem stems from the fact that species' niches may be noncomparable, that is, lying on different dimensions or having different *numbers* of dimensions. Just as one cannot speak sensibly about the total volume of a structure consisting of a plane, a cube and a line, it does not make sense to speak of a communal hypervolume in Euclidean space when constituent species have niches of different dimensionalities (i.e., different measures). The communal hypervolume idea is also not a fruitful one for gaining a sense of the geometry in the overlap structure of the community. Species niches that overlap in some dimensions but not in others have disconnected hypervolumes, giving a trivial topology to the communal niche. The only nontrivial connected regions consist of those species niches that overlap on, and are *completely* represented by, a common set of resource axes. The clear geometric intuition that made Hutchinson's characterization so attractive is therefore lost in the communal hyperspace model, and the hypervolume concept as it is rooted in a Euclidean metric space unfortunately does not work at the larger community level.

2.2 HANDLING THE DIMENSIONALITY PROBLEM

We will now discuss some solutions to the problem of representing the mathematical structure of species interrelationships in a community.

By *community* we mean a group of interacting species in a homogeneous or well-defined habitat such that every species in the ensemble overlaps with at least one other species on some set of resources or niche dimensions. Thus *community* here corresponds to the connected network of direct competitive interrelationships. A species whose niche overlaps no other species forms a trivial or isolated community. The motivation here is to obtain a natural definition that parallels the Hutchinson Whittaker idea but that generates a sensible community topology. Thus, all species in a community must be related through some sequence of niche overlaps, and this transitive chain forms the network of mutual interdependence among species.

To sharpen the discussion we will examine a community of estuarine fishes studied by Day (1967) within the Knysna Estuary near Cape Province, South Africa. The niche axes in this example, as in many of the cases considered in the following chapters, are trophic dimensions of a binary nature: a species either uses the resource or it does not. Cohen (1978) gives a good case that trophic factors are the meaningful niche dimensions for studying ecological interdependence, as does Hurlbert (1980) in his discussion of the niche as a resource set.

Figure 2.1 presents the food web matrix for this community where the columns represent consumer species (members of the hypervolume) and the rows represent resources (niche axes). The term "species" will be used loosely here to denote trophic equivalence (see section 2.3 for a complete discussion). The 1's and 0's in the matrix are used to indicate whether or not a consumer utilizes a particular resource. Consumer species 12 (*Johnius hololepidotus*), for example, can be seen to use the resources 6 (*Mugil*, a teleost) and 7 (*Upogebia*, a crustacean). *Rhabdosargus* (14) illustrates the fact that a species may be both a member of the hypervolume (column) and one of its axes (row), that is, a species may be both predator and prey. In total, 10 resources are shown here, meaning that the assemblage is defined in a 10-dimensional niche space. Notice that because there are no isolated species or disconnected species groups, this represents a single community.

2.2.1 The Consumer Overlap Graph

The *niche overlap graph* for consumers is a simple representation of direct competitive relationships. This non Euclidean picture of the communal niche was recently introduced by Cohen (1978), and portrays the network of niche overlaps as an undirected graph. Here a graph is defined as a finite non-empty set V of vertices (points),

Community Food Web Matrix (Knysna Estuary)

	12	13	14	15
2			I	
5				I
6	I			I
7	I	I		
8		I	I	
9		I	I	
10		I		
11		I	I	
13				I
14				I

LEGEND

2. Attached plants
5. Hyporhamphus
6. Mugil
7. Upogebia
8. Lamya
9. Solen
10. Arenicola
11. Hymenosoma
12. Johnius
13. Lithognathus
14. Rhabdosargus
15. Hypacanthus

Figure 2.1 Food web matrix for the Knysna Estuary (Day, 1967). Columns represent consumer species and rows represent the resources they use.

together with a prescribed set E of undirected edges (lines) connecting any subset of vertex pairs. Consumers in the niche overlap graph are represented by vertices, and two vertices are joined by an edge if and only if the consumers share a common resource.

Figure 2.2 contains an illustration of a niche overlap graph for the estuarine fish community constructed from the food web matrix in Figure 2.1. Again, two consumers are joined by an edge if they share a common resource i.e., if their columns in the food web matrix overlap. For example, the two predatory fish, *Hypacanthus amia* (15) and *Johnius hololepidotus* (12) are joined by an edge because they overlap in the utilization of a common fish prey, *Mugil* sp. (6). Again, the consumer overlap graph presents a picture of the potential competitive interactions structuring the community.

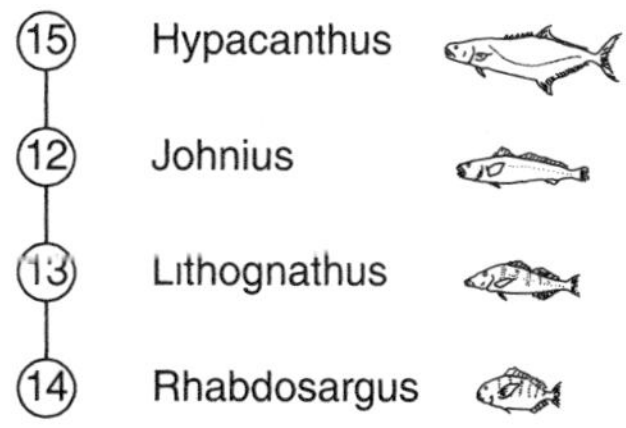

Figure 2.2 Niche overlap graph $G_R(C;\lambda)$ for Knysna Estuary (Day, 1967). Vertices represent species and edges indicate shared resources.

In graph theoretical terms, the niche overlap graph is an example of an *intersection graph* or a graph whose points correspond to sets (consumers are sets of resources) and whose lines correspond to the pairwise non-empty intersection of these sets. The mathematical structure of the consumer overlap graph has consumer species as *objects* and "non-empty intersection in resource use" as the logical *relation* among objects. More formally, this relation λ is defined on the product space of consumers C and resources R (i.e., $\lambda \subset C \times R$), thus it is natural to denote the consumer graph by:

$$G_R(R; \lambda)$$

which is similar to the simplex-complex notation used in algebraic topology. Notice that a *connected component* in $G_R(R; \lambda)$ corresponds exactly with our definition of community in that every point is connected to every other point through a sequence of edges.

The niche overlap graph has some distinct advantages as a mathematical representation of the community niche. It is well defined, it provides a clear geometrical sense of the competitive network in the community, and as demonstrated by Cohen (1978) it addresses the question of the minimum dimensionality required to represent overlaps in niche space. However it pales as a metaphorical device when compared with the richness and detail of the Hutchinson-Whittaker concept. In particular, the picture of overlap among species as presented in $G_R(R; \lambda)$, with overlap reduced to a single edge, cannot address the potential multidimensionality of overlaps—the sense that many resources can be involved in any single overlap. For example, in Figure 2.2 although the predatory fish *Rhabdosargus* (14) and *Lithognathus* (13) share three common resources (8, 9 and 11), their relationship is represented in $G_R(C;\lambda)$ as a single unspecified linkage. In addition to losing information about the number of dimensions involved in each overlap and their identity as resources, we also lose the multidimensional character of species niches themselves as they are collapsed into single undifferentiated points. It would be useful to have a fuller characterization of the niche that preserves more of the richness of the Hutchinson-Whittaker concept, but that is well defined.

2.2.2 The Resource Graph: A Tinker Toy Model

Recall that consumers in $G_R(C;\lambda)$ are the objects (points) and resources determine the relations (links). A richer and more expanded view is possible by simply inverting the picture: making resources the points and having consumer species define the relations among these points. The resulting inside-out construction or *resource graph* is similar in concept to the hypervolume idea in that species are viewed as multidimensional meta-objects in a resource space.

In the consumer graph the relation λ is defined on the product space of consumers and resources (i.e., $\lambda \subset C \times R$). We now define the resource graph to be the formal conjugate of the consumer graph, denoted by $G_C(R;\lambda^{-1})$, where $\lambda^{-1}\ R \times C$. Whereas consumer species are linked if they share a common resource, in $G_C(R;\lambda^{-1})$ resources are linked when used by a common consumer. Therefore a consumer is a meta-object represented as a *complete* subgraph or a *clique* in the resource graph—a subset of vertices that are maximally connected so that all vertices in {S} are linked to all others ({S} being the set of resources that a species uses). The justification for this will be made plain in the following example.

Consider again, the food web matrix for the estuarine fish community in Figure 2.1. To construct the resource graph from this matrix, read down the *columns* and link all resources that fall within each column, i.e., that are used by a given consumer. Thus resources are related (linked) if they belong to the same consumer species, with edges indicating relationship to a common consumer so that a species niche is defined as the maximal *clique* over the resources it utilizes. Figure 2.3 shows the resource clique for each consumer in

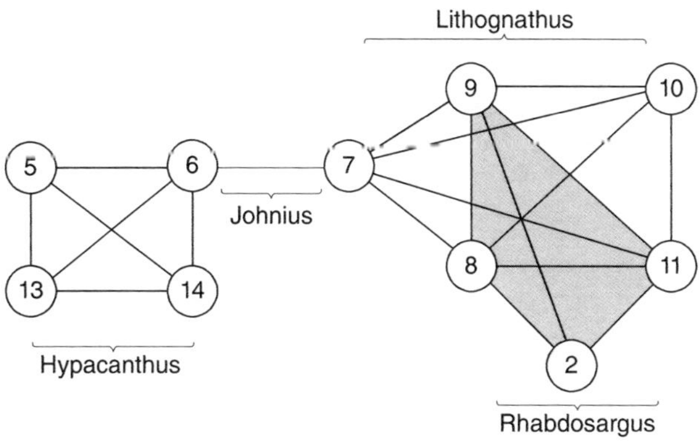

Figure 2.3 Resource graph $G_C(R;\lambda^{-1})$ for Knysna Estuary (Day, 1967). Vertices are resources and edges indicate use by a common consumer. A consumer is therefore represented as a clique over the resources it utilizes.

this community. Observe that these cliques are the graphical representation of the n-dimensional niche for each species, where n is the number of vertices in each clique. Each niche is *convex* because every vertex is connected by a straight line segment to every other vertex. The final graph for the community $G_C(R;\lambda^{-1})$ is the union of the labeled resourcc cliques (Figure 2.3).

Figure 2.4 (Part B) illustrates another example of a resource graph for an aquatic community studied by Qazim (1970) in the Cochin backwater, India. Compare this picture of this niche the overlap graph in part C of this figure.

Clearly, $G_C(R;\lambda^{-1})$ contains more information than does $G_R(C;\lambda)$, and is similar to the Hutchinson-Whittaker concept in that it provides a multidimensional portrait of the communal niche. Here, however, instead of the resource *axes* we are given resource *points* and in place of hypervolumes we have n-dimensional niche *cliques.*

For pedagogical reasons it may be desirable to draw the cliques for each species in $G_R(C;\lambda)$ in a different color. This is especially helpful when one species niche is properly contained in another. The resulting community resource graph from this procedure is an example of a multi-graph, or a graph having multiple edges.

Additionally, a three-dimensional analogue can be constructed with tinker toys using struts of a given color to represent a particular species

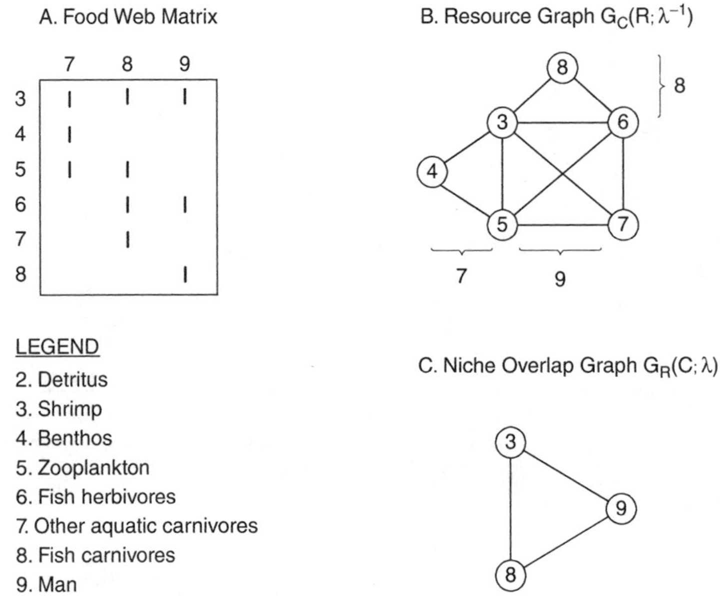

Figure 2.4 Food web matrix, niche overlap graph $G_R(C;\lambda)$ and resource graph $G_C(R;\lambda^{-1})$ for Cochin backwater (Qazim, 1970). $G_C(R;\lambda^{-1})$ preserves more detail than $G_R(C;\lambda)$.

and balls of modeling clay to represent resources. That it should always be possible to construct the multigraph version of $G_C(R;\lambda^{-1})$ in tinker toys follows from a well-known theorem showing that any graph can be *embedded* (no edges crossing) in R^3 (Harary, 1972).

2.2.3 Species as Convex Polyhedra

A more abstract yet analytically powerful characterization of the resource graph is possible with algebraic topology. Here, the notions of *clique* and *graph* are replaced by the terms *simplex* and *complex* as defined below.

Let $v^o, \ldots v^n$ be vectors R^n. A vector V is said to be *affinely dependent* on them if there exist real numbers $\lambda_o, \ldots, \lambda_n$ called *barycentric coordinates* such that:

$$\sum_{i=0}^{n} \lambda_i = 1 \text{ and } V = \lambda_0 v^0 + \ldots + \lambda_n v^n$$

Suppose further that $v^o, \ldots, v^n$ are affinely independent (none affinely dependent on the rest). Then, a *simplex* (closed) having vertices $v^o, \ldots, v^n$ is defined to be the set of points affinely dependent on $v^o, \ldots, v^n$ such that every barycentric coordinate is ≥ 0. The *boundary* of the simplex consists of those points that have at least one barycentric coordinate equal to 0, and its *dimension* n is simply the number of vertices minus 1. Thus an n-dimensional niche in Hutchinson's parlance corresponds to an n – 1 dimensional simplex over {S}.

Figure 2.5 illustrates this definition of simplex with some examples. The isolated vertex is the 0-dimensional or 0-simplex, two points joined by an edge form the 1-simplex, three vertices define

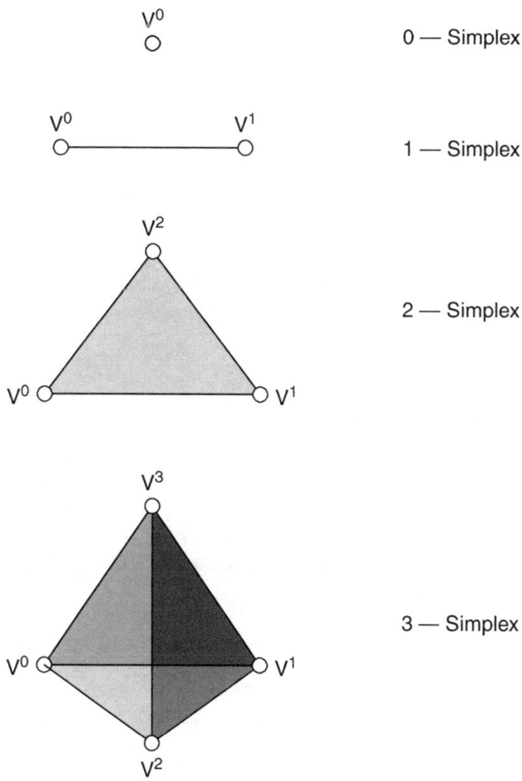

Figure 2.5 Examples of *n*-dimensional simplexes, i.e., *n*-simplexes. The dimensionality of a simplex is 1 minus the number of vertices, i.e., 1 minus clique dimension.

the 2-simplex, and the solid tetrahedron is the 3-simplex. Notice how the notion of simplex corresponds with the graphical term "clique". Both are similar in that vertices and edges in the simplex identify with vertices and edges in the clique, however in dimensions above 1 the simplex characteristics of interior, exterior and boundary, as defined by the barycentric coordinates, differ from those in the graph. The 2-simplex, for example, is the entire shaded triangle (Figure 2.5), its boundary consists of edges and its interior is the shaded region within. By contrast, the corresponding graph consists only of the vertices and edges of the simplex, Harary (1972) in fact defines a graph as a complex of simplexes of dimension 0 or 1. A simplex, therefore, can be viewed as a clique in a higher dimensional space with an interior that is filled. Conversely a clique can be viewed as the vertices and edges of a simplex, or the *1-skeleton* of a simplex.

Notice that because the clique is convex, so is the simplex. That is, if u and w are two vertices of the simplex then every straight line segment joining u and w belongs to the simplex. Indeed the simplex is the smallest convex set containing its vertices, which is expressed by saying it is the *convex hull* of its vertices.

In correspondence with our graphical representation of the species niche as an n-pointed clique it is also possible to represent the n-dimensional niche of a species as a simplex over n resource vertices. Represented as a simplex, therefore, the niche of a species can be visualized as a convex polyhedron on n vertices. Again, the dimensionality of this polyhedron is by convention $n - 1$.

Building now to the larger picture, as in $G_C(R;\lambda^{-1})$, we simply glue the individual species polyhedra together at the appropriate matching vertices to generate a portrait of the communal niche resembling a multidimensional crystal. This complex of species simplexes is a *simplicial complex*, denoted $K_C(R;\lambda^{-1})$. The simplicial complex $K_C(R;\lambda^{-1})$ is the relational equivalent of the resource graph insofar as λ^{-1} serves the same function in K as in G. $G_C(R;\lambda^{-1})$ can be said to be the 1-skeleton of $K_C(R;\lambda^{-1})$ and consists of the vertices and edges of K. The dimension of the complex is defined by convention as the dimension of the largest simplex, and the *underlying space* {K} is the totality of distinct resource vertices supporting the community.

In Figure 2.6 we apply the simplex characterization to the example from Figure 2.4 to obtain a picture of the communal niche that looks

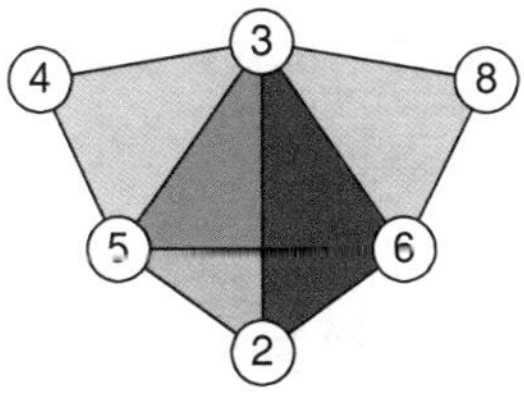

Figure 2.6 The Knysna community (Day, 1967) represented as a simplicial complex $K_C(R;\lambda^{-1})$. As in $G_C(R;\lambda^{-1})$ the vertices correspond to resources and maximal simplexes represent species.

like a tetrahedron with triangular wings. As will be seen in the following chapter, the motivation for introducing the simplicial complex model derives from certain geometric properties of this representation. Atkin (1971, 1972) and Casti (1979) have been early advocates of the potential value of simplicial complexes, used in analogous ways for understanding various applied problems.

The principle drawback in representing the community niche as a simplicial complex is that it is not possible to draw polyhedra of dimension greater than 3 (simplex dimension) on a piece of paper. Looking back at the estuarine fish community of Figure 2.2 we see that one of the species (13, *Lithognathus*) has a niche consisting of 5 vertices. This species is therefore a 4-simplex and the community complex cannot be represented as a neat perspective drawing as was done for the ensemble of Figure 2.6. A conservative upper bound on the dimensionality of a Euclidean space needed to realize (embed) a simplicial complex of dimension n is 2n + 1 (Giblin, 1977).

2.3 TROPHIC EQUIVALENCE

Because the models discussed so far require that resources be discrete, it is useful to have an operational method for differentiating resources which fall most naturally along a continuous spectrum. A simple recipe that does this is illustrated in Figure 2.7 where a continuous one-dimensional resource spectrum is broken down into discrete regions according to how it is used by consumers. The principle at work here is that it is important to discriminate between resources only insofar as the resources discriminate between consumers. This will be true for the questions to be addressed here. Therefore, one can

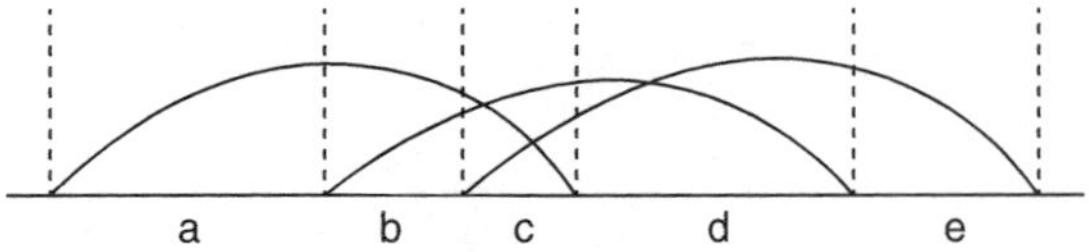

Figure 2.7 A continuous resource spectrum may be divided into discrete resources according to how it is utilized.

aggregate resources in a food web matrix whose column sums are equal and equal to their dot products without changing the essential structure of how consumers are interrelated in $G_R(C;\lambda)$, $G_C(R;\lambda^{-1})$ or $K_C(R;\lambda^{-1})$. The size of the cliques and the simplexes in $G_C(R;\lambda^{-1})$ and $K_C(R;\lambda^{-1})$ may be altered, but who interacts with whom and the homology of the complex will remain unchanged. Conversely, consumers which use identical resources may be aggregated without changing the non-multigraph version of $G_C(R;\lambda^{-1})$ or $K_C(R;\lambda^{-1})$. The principle in operation here is *trophic equivalence*: resources are different only if they distinguish consumers and consumers are different only if they distinguish resources.

SUMMARY

The Hutchinson-Whittaker model for the communal niche is mathematically ill-defined, and when taken literally, will lead to a trivial topology consisting of disconnected species niches in the community hyperspace. In response, three alternative characterizations are examined each of which may give rise to a more interesting architecture for the communal niche.

Cohen's niche overlap graph $G_R(C;\lambda)$ is reviewed and is shown to provide a picture of the potential competitive relationships in the community. Here two consumer species are linked if and only if they share a common resource. Although $G_R(C;\lambda)$ is analytically well defined, it lacks the nuance and richness of the full multidimensional portrait suggested by the Hutchinson-Whittaker concept.

An alternative characterization, the resource graph $G_C(R;\lambda^{-1})$, is proposed which is demonstrated to preserve the full multidimensional flavor of the hypervolume idea. It is seen as the conjugate or inside-out version of Cohen's niche overlap graph. For pedagogical purposes $G_C(R;\lambda^{-1})$ can be constructed as a tinker toy model with articulation points corresponding to resources, and species niches represented as struts of a given color maximally connecting resource points. This maximally connected set of resource points is exactly equivalent to an n-dimensional species niche. The community niche is then created by linking up the toy species niches so that their labeled articulation points (resources) match. The resulting tinker toy model provides a robust geometrical portrait of the multidimensional community niche.

Finally, we recast the concept of the resource graph so that species niches correspond to simplexes and the communal niche then becomes a simplicial complex, $K_C(R;\lambda^{-1})$. This gives rise to a solid tinker toy model where species niches are viewed as convex polyhedra and where the vertices of the polyhedral niches correspond to resources. Recasting $G_C(R;\lambda^{-1})$ in this way provides better physical intuition for the architecture of the community niche and, as will be demonstrated in the next chapter, suggests some clear questions about structure and organization in real niche spaces.

3

HOLES IN NICHE SPACE

One of the now classic problems in theoretical ecology concerns how similar competing species can be, and still persist in an equilibrium community. That is, how closely can species be packed and yet remain stable? Robust results taking us far with this problem have been established for packing in one-dimensional (MacArthur 1972, May 1974), multidimensional (May 1973, 1975) and stochastic environments (Turelli, 1976).

As the complement to the question of maximal packing one can also ask whether there are any *minimal* constraints to species packing. That is, while the classical species packing problem addresses the question of the dynamically constrained maximum density of species niches allowed, it is also useful to address the complementary topological problem and ask whether there are any restrictions on the minimum "density" actually realized in the environment.

As discussed in Chapter 2 this problem is only defined in the Hutchinson-Whittaker model if species occupy an identical set of niche axes. Species with a resource axis not in common will have niches that do not overlap in the community hyperspace. Accordingly there is no basis for applying the Hutchinson-Whittaker concept to the topological question of minimal packing. On the other hand, the resource graph $G_C(R; \lambda^{-1})$ and especially the simplicial complex $K_C(R; \lambda^{-1})$ models of the niche may be applied naturally to this problem.

Suppose we have a configuration in the communal niche $K_C(R; \lambda^{-1})$ as shown in Figure 3.1A, where each species is defined on three resources and where the species niches overlap to form a circuit or closed chain with a hole in the center. Compare this with the community portrait in 3.1B where the species niches combine without leaving a gap. The intuition conveyed by this simple comparison is that one may be able to take the presence or absence of holes in the underlying community resource-complex as the topological criterion for packing in the minimal sense. Here a community complex may be described as either having holes or as being *homologically dense*, or having a *trivial homology*. In the latter case the underlying space for the community {K} (defined as the totality of resources used), is filled in a conservative manner whereas in the former instance the niche is filled in a loose or convergent style. For example, species d of Figure 3.1A is joined to the parent complex by overlapping two species (a and c) which themselves do not overlap; furthermore it overlaps with respect to two resources (1 and 7) which appear to be dissimilar in that they do not belong to a single species or even to adjacent and hence similar species in the parent complex. As will be discussed in detail in section 3.2.3, similarity among resources is implied in their ownership by a single consumer species. For example, if simplexes a, b and c in Figure 3.1A represent three bird species each of which inhabits either the forest floor, the lower, or upper canopy, then the resources that a given species uses should be associated with its unique spatial location. The spatial location provides a *local ordering* to the resources.

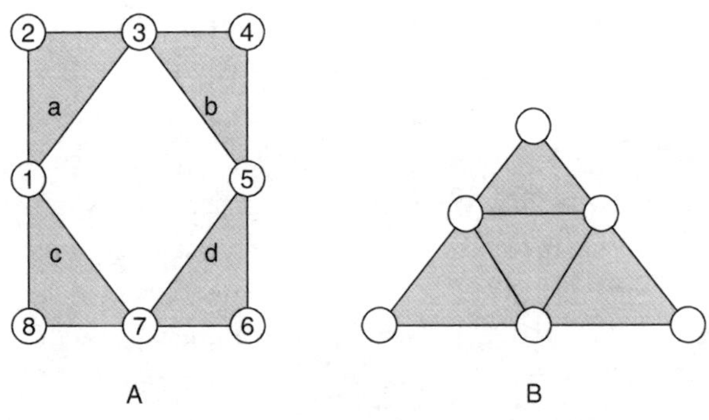

Figure 3.1 Two alternative packing schemes in $K_C(R; \lambda^{-1})$. In both cases all species are defined on 3 resources. In B, species are packed conservatively while in A they are packed loosely leaving a hole in the complex.

Thus, in Figure 3.1A species d uses resources on the ground and upper canopy but ignores resources in the middle elevations. Species d has a heterogeneous niche that ignores the ordering imposed by spatial location and participates nonconservatively leaving a gap in the middle elevations. If a, b, c and d are close evolutionary relatives then d would represent convergence in the character states of the niche. Hence as will be discussed more fully in section 3.2.3, the formation of a hole in the community complex represents nonconservative or convergent occupancy of the resource space whereas the absence of holes identifies with a conservative packing scheme that follows a local ordering in the resources. In simple terms, therefore, the question of minimal packing translates to asking whether the occupied resource space resembles a completed jigsaw puzzle, or a puzzle with gaps left by missing species—species who do not follow the local ordering of resources. Valentine (seminar) suggests an alternative geometric analogy to discuss maximal packing over a set of character states in evolution in terms of the tesselation of a resource plane. It is plausible that $G_C(R; \lambda^{-1})$ and $K_C(R; \lambda^{-1})$ could be applied to this problem, however for the moment we will restrict attention to the minimal packing question and ask whether the community resource set {K} is filled in a loose manner that produces holes like swiss cheese or whether its topology is homeomorphic to a solid monterey jack.

In a broad sense MacArthur anticipated the question of minimal packing in his Q-theorem papers (1969, 1970). He prophetically demonstrated that at competitive equilibrium the discrepancy between available production and resource utilization should be minimized. This minimum however differs from the one considered here in that it involves packing of *individuals* in a resource spectrum rather than packing of species in niche space.

It should be emphasized that holes are defined here in terms of the consumer species present and the resources that they use. Therefore, a hole in the community complex does not correspond to resources that are left completely unutilized i.e., disconnected vertices in the complex. However, as will be explained further in Chapter 5, it may be possible to view a hole as a vacancy or missing species in the ensemble.

3.1 THE MATHEMATICAL STRUCTURE OF HOLES

This section sketches the fuller algebraic details of the intuitive remarks on minimal packing made in the preceding paragraphs. It begins in 3.1.1 by outlining the elementary background for the standard homological view of holes in simplicial complexes and goes on in 3.1.2 to suggest a simple computational method for finding 1-dimensional holes in K from its 1-skeleton. Readers not interested in the technical basis of holes or how to compute them may wish to skip this section and instead look up the definition of Betti numbers β_n (K) in and study Figure 3.4 to get a clearer sense of what is meant by 1 and 2-dimensional holes.

3.1.1 Standard Homology for $K_C(R; \lambda^{-1})$

Material contained briefly here is covered in much greater detail in texts by Hilton and Wylie (1960), MacLane (1963), and Wallace (1970), and is patterned after Atkin's (1972, 1974) and Giblin's (1977) elegant and economical expositions of these ideas in terms of elementary abelian groups. It will focus on the homological properties of simplicial complexes or the study of multidimensional holes. The basic principle of this analysis involves using the algebraic properties of cycles in $K_C(R; \lambda^{-1})$ for gathering geometric information about the community complex.

Consider the case of a complex $K_C(R; \lambda^{-1})$ in which dim (K) = n. Such a complex may be said to consist of 0-simplexes, 1-simplexes, . . . , n-simplexes (see Figure 3.2). Now assume that we impose an *orientation* on K, induced by ordering the resource vertex set R to yield $(v^o, v^1, \ldots, v^{|K|})$,

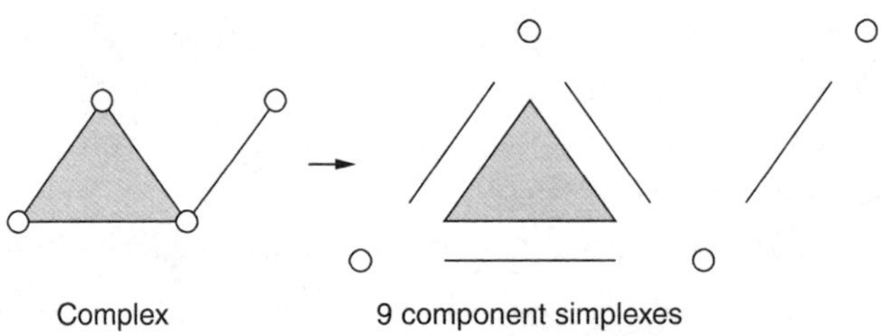

Figure 3.2 A simplicial complex and its component simplexes.

with |K|, once again, representing the cardinality of the underlying space, or total number of distinct resources in the community complex $K_C(R; \lambda^{-1})$. The particular ordering is immaterial here since most global properties of simplicial complexes (e.g., homology) are invariant under a change in orientation.

Let α_p be the number of p-simplexes of K. The so-called p[th] chain group of K, denoted $C_p(K)$, may now be constructed by taking the formal linear sum of these p-simplexes. The elements of $C_p(K)$ are called *p-chains* c_p, each p-chain having the form:

$$c_p = \lambda_1 \sigma_p^1 + \ldots + \lambda_{\alpha p} \sigma_p^{\alpha p} \tag{3.1}$$

with λ's integers, λ ε J. Two p-chains are added simply by adding corresponding coefficients as follows:

$$c_p + c_p^{'} = (\lambda_1 + \lambda_1^{'})\sigma_p^1 + \ldots + (\lambda_{\alpha p} + \lambda_{\alpha p}^{'})\sigma_p^{\alpha p}$$

The totality of these p-chains taken together form the pth chain group $C_p(K)$ which is the free *abelian group* over the set:

$$\left\{\sigma_p^{'}, \ldots, \sigma_p^{\alpha p}\right\};$$

an abelian group being a set A containing the zero and inverse elements together with a binary operation "+" defined on A which is associative and transitive. This simple notion of $C_p(K)$ and its elemental p-chain enables us now to construct an algebraic definition of boundary which is essential for characterizing holes in $K_C(R; \lambda^{-1})$.

Recall from section 2.2.3 that the boundary of a 1-simplex is represented by its vertices and that of a 2-simplex by its edges. Similarly, with every p-chain C_p we now aim to associate a particular (p − 1)-chain as its boundary denoted by $\partial_p c_p$. It is possible to define $\partial_p c_p$ precisely in terms of the boundary of a simplex $\partial_p c_p$ where:

$$\partial_p c_p = \partial_p \left(\sum_i \lambda_i \sigma_p^i \right) = \sum_i \lambda_i \partial_p^i \sigma_p$$

for 0 ≤ p ≤ n. That is, we require that ∂_p be a simple *homomorphism* (preserves the additive structure) from $C_p(K)$ into $C_{p-1}(K)$.

$$\partial = \partial_p : C_p(K) \rightarrow C_{p-1}(K)$$

In particular if $\sigma_p = (v^0, \ldots, v^p)$ is an oriented p-simplex of K, where $p > 0$, then the *boundary* of σ_p is the (p – 1)-chain.

$$\partial_p \sigma_p = \sum_{i=0}^{p} (-1)^i (v^0 \ldots \tilde{v}^i \ldots v^p) \tag{3.2}$$

where the tilde over v^i indicates that vertex i is omitted.

To illustrate the above definition of boundary consider the arbitrarily ordered 2-simplex in Figure 3.3, $\sigma_2 = (v^0\, v^1\, v^2)$. The arrows here indicate the induced orientations on the edges. Upon applying equation (3.2) it is seen that:

$$\begin{aligned}\partial_2 \sigma_2 &= \partial_2 (v^0\, v^1\, v^2) \\ &= (-1)^2 (v^1\, v^2) + (-1)^3 (v^0\, v^2) + (-1)^4 (v^0\, v^1)\end{aligned}$$

so that:

$$\partial_2 \sigma_2 = \sigma_1^1 - \sigma_1^2 + \sigma_1^3$$

which is a 1-chain and an element of $C_1(K)$. It should be clear that the boundary of a p-chain can be viewed as its image under the operator ∂, denoted variously by Im (∂) or $B_{p-1}(K)$, and that *every boundary is a cycle.* The converse, however, that *every cycle bounds* is not in general true and forms the basis for characterizing holes in a complex.

Notice that ∂ is not only a homomorphism but that it is also a *nilpotent* operator. That is, ∂ twice composed is the zero map:

$$\partial(\partial c_p) = \partial^2 c_p = 0$$

In the example in Figure 3.3, for instance, one sees that:

$$\begin{aligned}\partial^2 \sigma_2 &= \partial_1 (\sigma_1^1 - \sigma_1^2 + \sigma_1^3) \\ &= \partial(v^1\, v^2) - {}_1\partial(v^0\, v^2)_1 + \partial(v^0\, v^1) \\ &= \left[(v^2) - (v^1)\right] - \left[(v^2) - (v^0)\right] + \left[(v^1) - (v^0)\right] \\ &= 0\end{aligned}$$

We may therefore, characterize the image of $C_p(K)$ under ∂, that is the boundary subgroup $B_{p-1}(K)$, in terms of nilpotence:

$$\partial B_{p-1} = 0_{p-2}$$

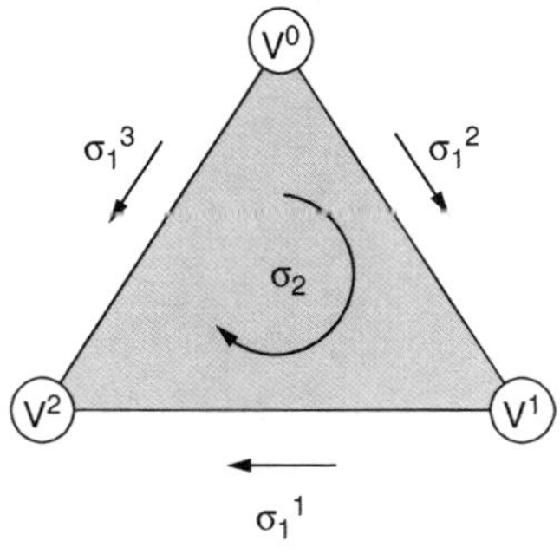

Figure 3.3 An oriented 2-simplex

alternatively $\partial[\mathrm{Im}(\partial)] = 0$. This is the algebraic version of "you can't peel a potato peeling."

It is significant that bounding cycles of C_{p-1} (K) (i.e., members of B_p) are not the only p-chains that vanish under ∂. Those p-chains for which $\partial c_p = 0$ are called *p-cycles,* and form the kernel of the homomorphism ∂. All p-cycles taken together comprise a subgroup of $C_p(K)$ denoted $Z_p(K)$, and the family of bounding cycles $B_p(K)$ (i.e., ∂C_{p+1}), in turn, form a subgroup of $Z_p(K)$. Hence,

$$B_p(K) \subset Z_p(K) \subset C_p(K)$$

The members of $Z_p(K)$ (i.e., p-cycles = z_p) which do not belong to $B_p(K)$ may be identified as representatives of the elements of the quotient group.

$$H_p(K) = \frac{Z_p(K)}{B_p(K)}$$

for p = 0, 1, . . ., n, called the pth *homology group* of K. The elements of this group have the form $z_p + B_p$ where B_p acts as the additive identity. As will be seen, the significance of $H_p(K)$ is that it measures the extent to which K has non-bounding p-cycles; that is, the extent to which K has "p-dimensional holes".

In the case that $H_p(K) = 0$ there is only one element in the quotient group and this is exactly the bounding cycle or B_p. Here, every element of $Z_p(K)$ is contained in $B_p(K)$, which is to say, every cycle is a bounding cycle. When $H_p(K) \neq 0$ there is more than one element in the quotient group $Z_p(K)/B_p(K)$, and hence, there must be at least one cycle at the pth level which is not a bounding cycle. This indicates the existence of a p-dimensional hole.

In Figure 3.3, for example, the only 1-cycle is the sum $\sigma_1^1 - \sigma_1^2 + \sigma_1^3$ (and multiples thereof) which is exactly the boundary $\partial_2\sigma_2$. Therefore, $H_1(K) = 0$. Because there is no $C_3(K)$ clearly $Z_2(K)$ is also empty, and $H_2(K) = 0$. However, at the 0th level one may show that $H_0(K) \neq 0$. Here the 0-chain may be written following 3.1 as:

$$c_0 = \lambda_1(v^0) + \lambda_2(v^1) + \lambda_3(v^2)$$

By taking the boundary of a vertex to be zero, it follows that c_0 is a 0-cycle. These 0-simplexes (v^0), (v^1), (v^2), however, are not independent in that they are arc-wise connected and can be expressed as:

$$(v^1) = (v^0) + \partial_0 c_1$$
$$(v^2) = (v^0) + \partial_0 c_1'$$

Here the vertex v^0 acts a special fundamental 0-cycle, denoted $\tilde{z}_0$, and all possible 0-cycles can thereby be generated from it by writing:

$$z_0 = \lambda\tilde{z}_0 + \partial_0 \text{ (some 1-chain)} \tag{3.3}$$

Notice that $\tilde{z}_0$ cannot be the boundary of any 1-chain, since clearly it is only a single point. Hence $\tilde{z}_0 \not\subset B_0(K)$ and $H_0(K) \neq 0$. $\tilde{z}_0$ is the single generator of $H_0(K)$ which is to say that it forms the basis of the 0th homology group in this example.

The number of independent generators of $H_p(K)$ is equal to the pth order *Betti number* $\beta_p(K)$ and corresponds to the number of *p-dimensional holes* in the complex. $\beta_p(K)$ is calculated as the rank of the pth homology group:

$$\beta_p(K) = \text{rank } H_p(K)$$

which again identifies with the number of independent generators of $H_p(K)$. In Figure 3.3 we saw at the 0th level that there was only a single generator $\tilde{z}_0$, and so here $\beta_p(K) = 1$. At the 1st and 2nd levels there were no generators (i.e., $H_1(K) = H_2(K) = 0$) and $\beta_1 = \beta_2 = 0$.

A more geometric way to look at generators of $H_p(K)$ is to see that they correspond to independent non-bounding p-cycles (recall $H_p(K)$

$= Z_p(K)/B_p(K)$). Therefore, $\beta_p(K)$ is exactly *the number of independent non-bounding p-cycles* in the complex. To see how this, in turn, corresponds to p-dimensional holes consider altering the 2-simplex in Figure 3.3 with a pair of scissors by cutting out the interior shaded region, leaving only the edges. The resulting structure defines a 3-clique and is a 1-complex consisting of the cycle over v^0, v^1, v^2 (Figure 3.4A). Notice that this cycle is a non-bounding cycle, the σ_2 having been removed. This leaves us with a single generator in the shape of $\sigma_1^1 - \sigma_1^2 + \sigma_1^3$ which is not the boundary of σ_2 and hence $\beta_1(K) = 1$. The single generator of $H_1(K)$ represents the presence in K of a hole bounded by 1-simplexes which shall be called a *1-dimensional hole.*

The 1-complex in Figure 3.4B consists of two hollowed out triangles. It is clear from the above argument that this structure possesses two non-bounding 1-cycles, and so here $\beta_1(K) = 2$. The first Betti number for the hollowed-out torus in Figure 3.4C is also 2, with the two independent classes of non-bounding cycles traversing the hub and girth of the torus respectively. Similarly a complex possessing a single spherical hole (bounded by the surface of a sphere as in Figure 3.4D) has an $H_2(K)$ which contains the single generator $\tilde{z}_2 \varepsilon B_2$, so here $\beta_2(K) = 1$, indicating the presence of a single 2-dimensional hole.

To summarize, geometric information about a complex may be gained from the algebraic structure of its cycles. In particular the

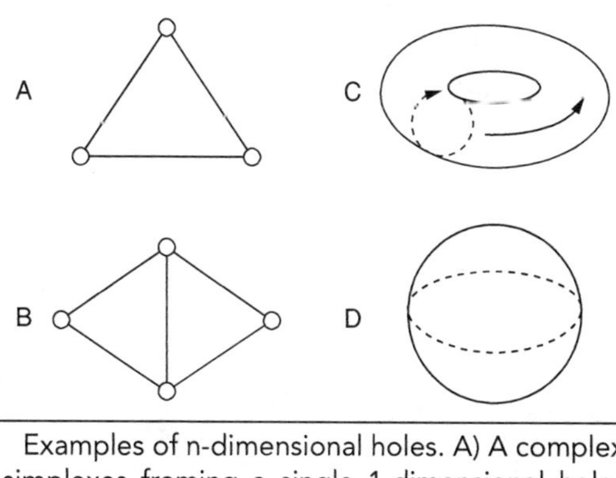

Figure 3.4 Examples of n-dimensional holes. A) A complex composed of three 1-simplexes framing a single 1-dimensional hole, $\beta_1 = 1$. B) A complex consisting of five 1-simplexes framing two 1-dimensional holes, $\beta_1 = 2$. C) A hollow torus containing two independent nonbounding cycles, $\beta_1 = 2$. D) A hollow sphere containing a single 2-dimensional hole, $\beta_2 = 1$.

p^{th} order Betti number $\beta_p(K)$ is relevant to the question of minimal packing in communities because it measures the number of p-dimensional holes in the underlying resource niche space. Whereas $\beta_0(K)$ is uninteresting in this context, in that for a community defined as a connected component $\beta_0(K)$ must always = 1, in higher dimensions, for $p \geq 1$, $\beta_p(K)$ can determine the extent to which the underlying niche space looks like a multidimensional swiss cheese or a solid monterey jack.

3.1.2 Computing $\beta_1(K)$ from $G_c(R; \lambda^{-1})$

As will be seen in the next section, no examples of the real community complexes studies here will have geometries containing holes of dimension greater than 1. It is useful in this event to have a simple operational method for calculating $\beta_1(K)$. The procedure proposed here makes use of the cycle structure of the resource graph $G_c(R; \lambda^{-1})$.

Recall Harary's (1972) definition of a graph as a complex of dimension 1 or 0. All 1-cycles in a graph are therefore non-bounding (no σ_2). It follows that the independent generators of $H_1(G)$ identify with the *independent cycles* in the graph. These taken together from the *cycle bases* of G defined as a minimal collection of cycles upon which all others depend (Harary, 1972). In Figure 3.5A, for example, the complete graph on 4 points has 3 independent cycles and in 3.5B the cycle basis consists of 4 independent cycles.

The first order Betti number for G is simply the dimension of the cycle basis or cycle rank.

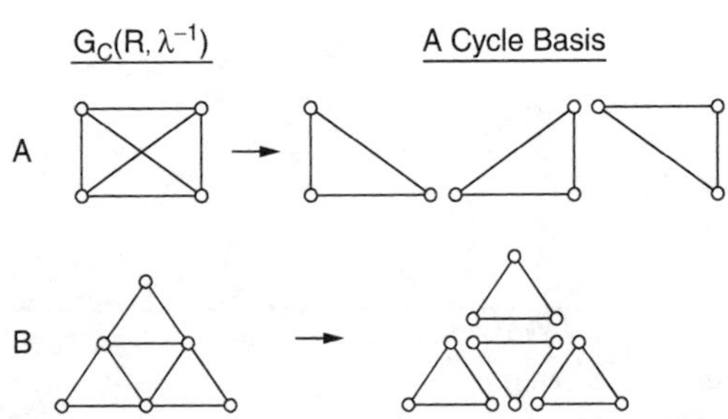

Figure 3.5 Two graphs and their respective cycle basis.

$$\text{rank } \beta_1(G) = Z_1(G)$$

Hence, the number of independent cycles in the graph describes the number of 1-dimensional holes in G.

A useful result for calculating $\beta_1(G)$ may be derived from the famous Euler-Poincare equation:

$$\alpha_0 - \alpha_1 + \alpha_2 - \ldots = \beta_0 - \beta_1 + \beta_2 - \ldots \tag{3.4}$$

where the β_n are the Betti numbers, and the α_n, as previously defined, are the numbers of simplexes of each dimension. Recall for a graph that β_0 is the number of connected components and that because no graph contains an n-simplex with n >, $\alpha_n = \beta_n = 0$ for all $n > 1$. Thus the Euler-Poincare equation reduces to:

$$\alpha_0 - \alpha_1 = \beta_0 - \beta_1$$

Furthermore for a connected graph where $\beta_0 = 1$, this relation may be rewritten as:

$$\beta_1(G) = E - V + 1 \tag{3.5}$$

where E is the number of edges and V the number of vertices in G. Hence, for $G_C(R; \lambda^{-1})$, $\beta_1(G)$ may conveniently be calculated as the number of edges minus the number of vertices plus 1.

We can extend this elegant result to compute $\beta_1(K)$ for the analogous complex $K_C(R; \lambda^{-1})$, when there are no higher dimensional holes. Recall that a one-dimensional hole in the community complex corresponds to an independent 1-cycle that does not bound some simplex. This is illustrated in Figure 3.6A where the independent 1-cycle 1, 2, 3 is not in the boundary of any simplex. It is a cycle which is not contained in a species clique or the 1-skeleton of a species simplex (see Figure 3.6A). It follows that one can compute $\beta_1(K)$ from $G_C(R; \lambda^{-1})$ simply by finding the number of independent cycles in G and subtracting the number of those that are bounding in K, i.e., that belong to the 1-skeleton of a species simplex. Therefore,

$$\beta_1(K) = \beta_1(G) - \text{rank } \beta_1(K) \tag{3.6}$$

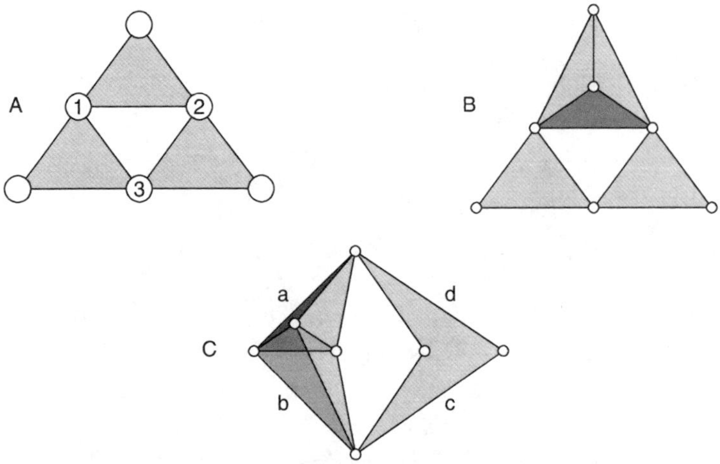

Figure 3.6 Three complexes $K_C(R;\lambda^{-1})$ containing 1-dimensional holes (non-bounding 1-cycles).

or $\beta_1(K)$ = rank $Z_1(G)$ – rank $B_1(K)$. The number of independent cycles is determined from the union of independent cycles for the species cliques in G corresponding to *maximal simplexes* (contained in no larger simplex) in K. For example, in Figure 3.6B, the maximal simplexes are the 2 shaded triangles and the solid tetrahedron. Using equation 3.5 on the cliques corresponding to each maximal simplex (1-skeleton) the number of independent bounding 1-cycles is easily computed as:

$$\text{Rank } B_1(K) = (3-3+1)+(3-3+1)+(6-4+1)=5$$

The total number of independent cycles is:

$$\beta_1(G) = 12-7+1 = 6$$

and applying equation 3.6 we find:

$$\beta_1(K) = 6-5 = 1$$

This indicates the presence of a single 1-dimensional hole in K. For simplexes that are joined at 3 or more points and whose intersection thereby contains cycles, rank $B_1(K)$ is computed as the *union* of bounding cycles for each simplex, calculated by working the inclusion-exclusion principle in the usual way to avoid counting cycles more than once. For the complex in Figure 3.6C, which consists of 2 tetrahedra and 2 triangles, the bounding cycles may be calculated as:

$$\text{Rank } B_1(K) = (3)_a + (3-1)_b + (1)_c + (1)_d = 7$$

where 1 is subtracted from b to account for the shared cycle in the intersection of simplexes a and b. In this example, $\beta_1(G) = 8$ so that $\beta_1(K) = 1$, again demonstrating the existence of a single 1-dimensional hole in the complex.

We can now apply these tools to data from real communities to locate and interpret weak spots or holes in the niche.

3.2 HOMOLOGY OF REAL FOOD WEBS

3.2.1 Results

From a broad survey of the literature, Briand (1982) assembled an excellent compendium of 40 "community" food web matrices representing ecological ensembles within homogeneous habitats (sensu Cohen, 1978). Appendix A contains a list of minor corrections obtained by consulting the original articles and by spot checking with more detailed information on the biology of the particular species involved. In all, these 40 community webs represent 102 communities defined here as connected components. Of the 102 connected components, 73 are non-trivial and contain more than 1 species, 60 contain 3 or more species and 50 of the communities contain 4 or more species. Table 3.1 lists the non-trivial communities, giving the numbers of species in each community, the list of clique dimensions for each species (i.e., niche size or number of resources used), and the 1st order Betti number $\beta_1(K)$ for the community complex. Appendix B contains the niche overlap graphs $G_R(C;\lambda)$ and a modified multigraph version of the resource graphs $G_C(R;\lambda^{-1})$ for each of the communities above. A representative resource graph illustrating the architecture of the niche for Bird's (1930) Aspen Parkland community is presented in Figure 3.7 (web 26A).

Two striking results emerge. First, none of these community complexes have holes of dimension ≥ 2, bounded by spheres or

Table 3.1 Statistics on $K_C(R;\lambda^{-1})$ for 73 Real Communities

Web I.D.	Community	Clique Vector[1]	$\beta_1(K)$	Random Assembly $Pr(B_n(K) \neq 0)^2$ all $n > 1$
1A	Cochin Estuary	2, 2, 2, 2	0	0.28
1B		3, 3, 4	0	0.19
2A	Knysna Estuary	1 (4×), 2, 2, 2	0	0.17
2B		2, 4, 4, 5	0	0.31
3A	Long Island Estuary	1 (7×), 2, 2, 3	0	0.18
3B		2 (4×), 3, 3, 3	0	0.52
3C		1, 1	0	0
4A	California Salt Marsh	1, 1, 1, 2 (6×), 6	0	0.53
5A	Georgia Salt Marsh	1, 1, 4	0	0
5B		1, 3	0	0
6A	California Tidal Marsh	1(4×), 2(4×)	0	0.28
6B		1(5×), 2(4×), 3, 3, 4, 6	0	0.56
6C		1, 1	0	0
7A	Narraganset Bay	1(4×), 2(5×), 3, 3, 4	0	0.55
7B		1, 1, 1	0	0
8	Bissel Cove Marsh	1(4×), 2, 2, 2, 3, 3, 4, 5	0	0.57
9A	Lough Ine Rapids	1, 2, 2, 3, 5	0	0.31
9B		1, 1	0	0
10	Exposed Intertidal (New England)	2, 2	0	0
11A	Protected Intertidal (New England)	1, 1	0	0
11B		2, 2	0	0
12A	Exposed Intertidal (Washington)	1, 1, 1, 1	0	0
12B		2, 2, 2	0	0.17
12C		2, 7, 7	0	0.21
13A	Protected Intertidal	1, 1, 1, 1	0	0
13B		2, 3, 8, 8	0	0.29

Web I.D.	Community	Clique Vector[1]	$\beta_1(K)$	Random Assembly $Pr(B_n(K) \neq 0)$[2] all n > 1
13C		2, 2	0	0
14	Mangrove Swamp	1(7×), 2	0	0
15	Mangrove Swamp	1(5×)	0	0
16A	Pamlico River	1, 1, 2, 2, 3	0	0.18
16B		1, 1, 3, 3	0	0
17	Marshallese Reef	1(6×), 2, 2, 5, 7	0	0.30
18A	Kapingarmarangi Atoll	1(11×), 3, 4, 6, 10	0	0.31
18B		1, 1, 4	0	0
19A	Moosehead Lake	1(5×), 2(4×), 5, 5, 6	0	0.63
19B		1, 1	0	0
20A	Antarctic Pack Ice	1(6×), 2, 2, 2, 3(5×)		0.86
20B		1, 2	0	0
21	Ross Sea	2, 2, 2, 3, 4, 4, 4	0	0.61
22A	Bear Island	1(8×), 2(4×), 3 4(4×), 5, 5, 6, 6	1	0.92
22B			0	0
23	Canadian Prairie	1(6×), 2, 2, 2, 3(4×), 4	0	0.73
24A	Canadian Willow Forest	1, 2, 2, 3, 5	0	0.30
24B		1, 3, 3	0	0
25A	Canadian Aspen Forest	1(4×), 2, 4	0	0
25B		1(7×), 2, 2, 2, 3	0	0.28
25C		1, 2, 6	0	0
26	Aspen Parkland	1(9×), 2 (6×), 3(6×), 4, 7, 9	0	0.95
27A	Wytham Wood	1(6×), 2(4×), 3, 3, 4, 5, 6	0	0.89
27B		1, 1, 2	0	0
28A	New Zealand Salt-meadow	1(8×), 2	0	0
28B		1(5×), 2, 2, 3	0	0.18
28C		1(10×), 2, 2, 2	0	0.17
28D		1, 1, 2	0	0
28E		1, 2	0	0

Web I.D.	Community	Clique Vector[1]	$\beta_1(K)$	Random Assembly $Pr(B_n(K) \neq 0)^2$ all n > 1
29A	Arctic Seas	1, 1, 1, 2	0	0
29B		2	0	0
29C		1(4×), 2, 2, 2	0	0.17
29D		1, 1, 1, 2, 2, 5, 6	0	0.31
30	Antarctic Seas	1, 1, 1, 2, 2, 3, 3, 3, 4, 5, 5	0	0.83
31	Black Sea Epiplankton	1, 1, 2, 3, 4(4×), 6, 6, 8, 8	0	0.91
32	Black Sea Bathyplankton	1, 1, 2, 3, 4(4×), 6, 7, 8, 8	0	0.91
33A	Crocodile Creek	1(14×), 2(5×), 3, 3, 4, 4, 4, 6	0	0.94
33B		1, 2	0	0
34	River Clydah	1, 2, 3, 3, 4, 4, 5, 5	0	0.64
35	Morgan's Creek	2(5×), 4(4×), 5, 5	0	0.94
36	Mangrove Swamp	1(4×), 2(4×), 3, 3, 5, 5, 7	1	0.86
37A	California Sublittoral	1(5×), 2, 2, 2, 3, 4, 5, 5	0	0.59
37B		1, 1	0	0
37C		1, 2, 3, 3	0	0.18
38	Lake Nyasa Rocky Shore	1(12×), 2, 3, 3, 4(4×), 5(4×), 7, 7, 8, 8, 9	0	0.96
39A	Lake Nyasa Sandy Shore	1(19×), 2(8×), 3, 4, 4, 6, 6, 8, 10	0	0.96
39B		1, 2	0	0
40	Malaysian Rain Forest	1, 1, 2(5×), 3	0	0.43
		Totals	**2**	**22.06**
				z = 7.51

[1]Clique Vector = numbers of resources in the component species niches.

[2]Probability that $\beta_n(K) \neq 0$, all $n \geq 1$, for a random complex from observed clique vector.

[3]The number of species in each community is equal to the number of elements in the clique vector (listed explicitly in Table 4.1).

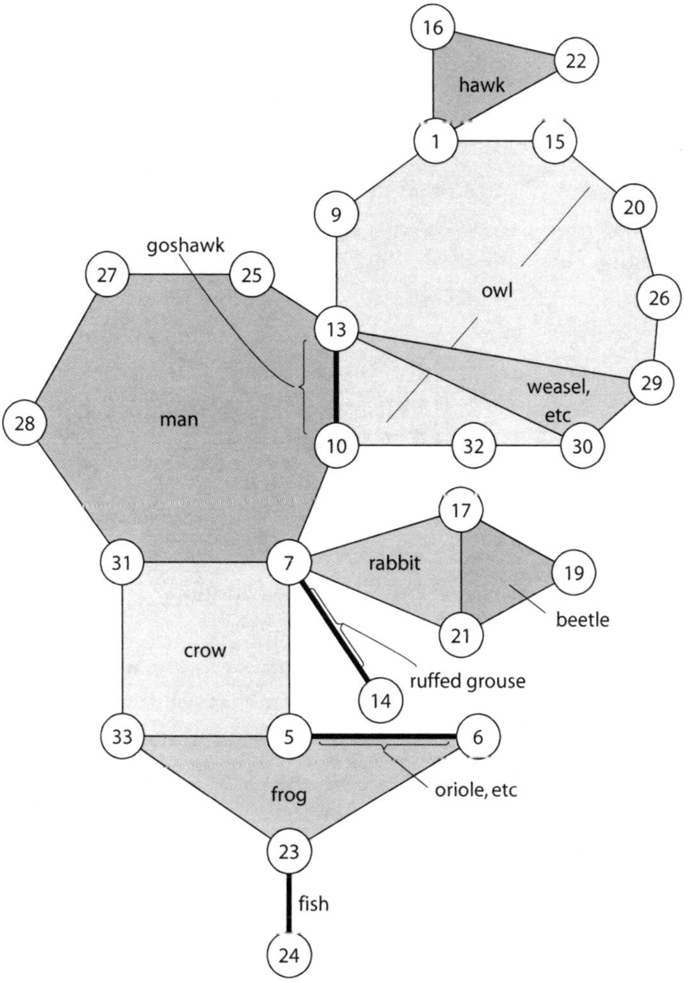

Figure 3.7 Bird's (1930) Aspen Parkland community. This figure contains a modified version of $G_C(R;\lambda^{-1})$ and $K_C(R;\lambda^{-1})$ where shaded polygons correspond to species.

hyperspheres. The geometries of real niches, although complicated in their richly detailed mosaic patterns, from the homological standpoint appear to be rather simple. Secondly, of the 60 communities that could possibly have 1-dimensional holes only two in fact do (webs 22A and 36A), and these as will be discussed in section 3.2.3 may be held with some reserve. The data here suggest, therefore, that holes appear to be extremely rare in real niche spaces and that natural communities emerge as being conservatively packed rather than loosely constructed.

3.2.2 Statistical Significance of Results

It is important to consider at this point whether these results are really as significant as they seem. In particular, can the rarity of holes be explained by a null model simply as a passive consequence of randomly fitting species niches together? Or does it reflect a genuine and active constraint operating on the assembly of communities?

To test this idea we will compute the expected probability that each community in Table 3.1 is free of holes when it is randomly assembled. Here we will use the observed species niches for each community (clique vector, Table 3.1) and combine them so that the order of incorporation of species, the number of resources overlapped (*s-cutset*) in each case, and the choice of which resources to overlap is determined at random with uniform distributions defined over appropriate limits. That is, the *s-cutset*, or the set of resources by which each additional species is tethered to the community, is determined completely at random. In effect this amounts to arbitrarily tossing the observed set of species niches on to the community complex so that each successive species overlaps randomly with those already present. Because zero overlap is not allowed the resulting complex is connected and conforms with our definition of community. A program that simulates this process was constructed in the C programming language (Kerrnigan and Ritchie, 1981). The last column of Table 3.1 gives the estimated probability that a given community is free of n-dimensional holes, $n \geq 1$, determined from 1000 random replicates. By summing these, we find that the expected number of communities possessing holes is 22.06. This compares poorly with the observed number of 2, and the difference is highly significant as indicated by a z value of 7.51 where:

$$z = \frac{\text{Obs} - \sum_i p_i}{\left[\sum_i p_i(1 - p_i)\right]^{1/2}}$$

Obs is the observed number of communities lacking holes and is the estimated probability that community i is free of holes. A z-value of 1.65 indicates significant difference at the 5% level (Snedecor and Cochran, 1967). Therefore, one must reject the null hypothesis that the absence of holes is a random artifact. If communities are randomly constructed one would expect many more holes than are observed. Thus, homological density in $K_C(R;\lambda^{-1})$ appears to be a real structural regularity.

To demonstrate that this is a reasonable null hypothesis we will compare the observed niche overlap measured by number of edges in $G_R(C;\lambda)$, to the overlap predicted. Following Cohen (1978) we want to show that our arbitrary null model is realistic in capturing, on average, the gross number of niche overlaps in a community. In this way we can begin to claim that the excess of holes predicted by the null model is due to *structure* in the observed niche overlaps and not simply to their *number*. It will be necessary here to translate $K_C(R;\lambda^{-1})$ into $G_R(C;\lambda)$ in the simulated communities to determine the predicted number of edges q in the corresponding niche overlap graph. As in Cohen (1978, p. 45) the distribution of P values is used to judge agreement between the observed and predicted q in $G_C(R;\lambda)$. If our null model is valid, we should expect the P values to be distributed uniformly in the interval 0 to 1. As witnessed in Figure 3.8 our simple null hypothesis is in reasonable agreement on this score and as such

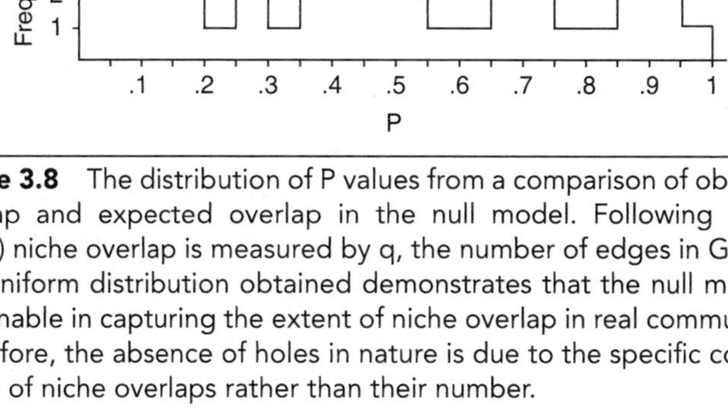

Figure 3.8 The distribution of P values from a comparison of observed overlap and expected overlap in the null model. Following Cohen (1978) niche overlap is measured by q, the number of edges in $G_R(C; \lambda)$. The uniform distribution obtained demonstrates that the null model is reasonable in capturing the extent of niche overlap in real communities. Therefore, the absence of holes in nature is due to the specific configuration of niche overlaps rather than their number.

may be a fair reference from which to judge the significance of the observed lack of holes in real communities.

These results demonstrate that the rarity of holes discovered in $K_C(R;\lambda^{-1})$ is significant and cannot be explained from a structure-free null hypothesis. Conversely, the rarity of holes demonstrates that communities are not arbitrary or random collections of species but have architectural constraints operating on $K_C(R;\lambda^{-1})$. This resonates with general arguments for the existence of organization and structure in natural communities (e.g., see Diamond, 1977) versus those claiming the lack of it (e.g., see Simberloff, 1978).

It should be mentioned that the results reported here are robust to the possibility that some of the studies may have failed to incorporate every species involved. It is obvious that if a subset of consumers lacks holes then the same subset will remain free of holes in the larger community complex. Indeed, the rarity of holes may be a conservative finding in that additional consumer species could act to fill holes in the subcomplex.

The fact that much of the data used here is coarsely aggregated into consumer groups and resource classes will also have no adverse effect on the presence or absence of holes in $K_C(R;\lambda^{-1})$. This follows if consumers and resources are aggregated according to *trophic equivalence* as presumably they were. This means that consumers may be lumped if they use the same resources (equal column sums and dot products in the food web matrix), and that resources may be combined if they do not differentiate consumers (equal row sums and dot products (cf. section 2.3)). Aggregating consumers in this way amounts to superimposing identical species cliques or simplexes on each other. This will have no effect on $G_C(R;\lambda^{-1})$ or $K_C(R;\lambda^{-1})$. Aggregating resources, which do not differentiate consumers, may reduce the number of distinct resources in species niches, but it should not systematically affect the presence or absence of holes in the complex.

On the other hand, completely arbitrary aggregation, without regard to trophic equivalence, may work in either direction. Combining consumer species into random groups may enhance the likelihood of creating holes in $K_C(R;\lambda^{-1})$, as in the example in the beginning of this chapter (Figure 3.1A) where a hole is generated by an arbitrary consumer group of birds that feed on the ground and ones that occupy the upper canopy. As will be discussed in the next section, these random or heterogeneous niches are more likely to form holes. Similarly, lumping resources may cause dissimilar consumer species to appear to overlap when in fact they do not, thereby creating an artificial hole. This may be weakly counterbalanced, however, by the effect that lumping resources might have on reducing the size of s-cutsets i.e., the number of resources by which each additional species is randomly attached to the community. Holes are only possible when there is an s-cutset that is greater than 1 in size.

At best then, in view of trophic equivalence, these possible inadequacies of the data should have little effect on these results. In fact, it seems likely that our finding showing holes to be rare in real community niche spaces is a conservative one.

3.2.3 Interpretation of Results

The appearance of a hole in the resource complex $K_C(R;\lambda^{-1})$ is related to the non-conservative addition of a species into the community and/or a species with heterogeneous niche that does not follow the local ordering of resources. To gain a deeper understanding of the apparent lack of holes in real communities we will now examine the two exceptional community complexes, each of which appears to contain a single 1-dimensional hole.

Web 22A (Figure 3.9) from Summerhayes and Elton's 1923 study of Bear Island, contains a hole which resembles a triangular flap held up by the tent pole. It is bounded by three consumer groups: A) skua and glaucous gull (19) whose simplex is defined on the 4 resources (15, 21, 26, and 27); B) northern eider, long tailed duck and red throated diver (21) (uses resources 23, 26, and 27); C) kittiwake, guillimots, fulmar petrel, little auk, puffin (20) (uses resource 15 and 23). It is significant that the 1-dimensional hole would vanish if the latter group, kittiwake etc. did not consume moss (15) or if in addition to moss it also took in either fresh water invertebrates (27) such as rotifers, tardigrades and oligochaetes, or if it ate diptera larvae (26). Both of these resource categories live in and around aquatic mossy environments. It seems plausible that the inevitability of ingesting these tiny invertebrates while feeding on moss or drinking water may be

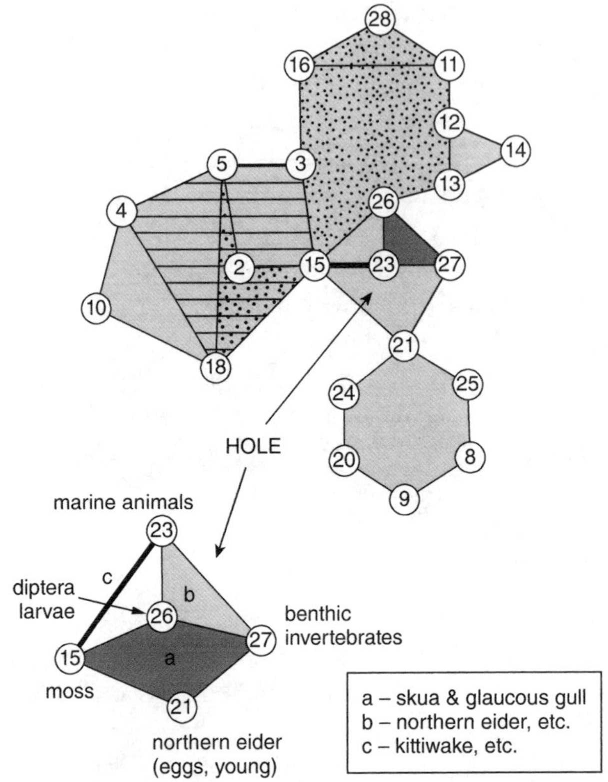

Figure 3.9 The Bear Island community (Summerhayes and Elton, 1923) contains a three-sided 1-dimensional hole.

as significant or insignificant as moss itself as a source of food. As it stands, however, this consumer group whose trophic niche is defined by the discontinuous categories, marine animals (23) and moss (15), appears on the community complex with a heterogeneous niche. The heterogeneity or hole in this case derives from the fact that omnivorous species that include moss in their diets, manage somehow to avoid eating fresh water invertebrates that are spatially associated with moss.

Walsh's (1967) study of a Hawaiian mangrove swamp, web 36 is the only other example of a community complex containing a hole. In this case the non-bounding cycle is defined by the resource points: (1) detritus, (2) diatoms, and (6) ostracods, (Figure 3.10). The

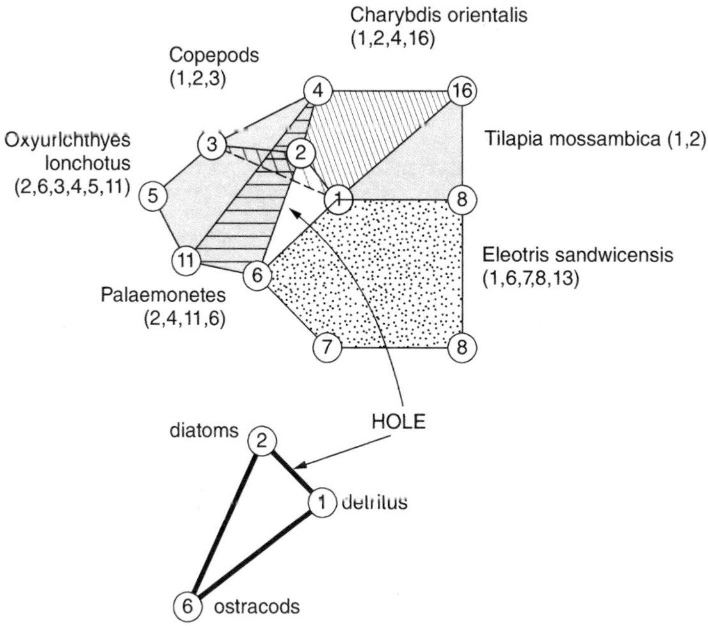

Figure 3.10 The community complex for Walsh's (1967) study of a Hawaiian mangrove swamp, web 36. A single 1-dimensional hole is bounded by the resource vertices (1) detritus, (2) diatoms, and (6) ostracods.

consumer species whose niches overlap these resources are illustrated in Figure 3.10. In every case only a pair of the resources are used by each consumer. It is worth noting that the gap which looks like the opening of a pup tent, would be filled if any one of the consumer species involved in framing it used the third resource. For example, the hole would dissappear if *Tilapia mossambica* or the portunid crab *charybdis orientalis*, which both eat diatoms and detritus, were to also include ostracods in their diets. Alternatively, if the goby species *Oxyurichthyes lonchotus* which eats a variety of yellow and green algae (2, 3, 4) as well as nematodes (5) copepods (11) and ostracods (6) took in detritus (hard to avoid when foraging on benthic creatures) the hole would also vanish. Similarly there is a good possibility that *Eleotris sandwicensis*, which feeds on ostracods and detritus, may close the gap by inadvertently ingesting benthic diatoms. Thus it is plausible that the lacunae uncovered in this community complex may in fact be filled. The hole or discontinuity in this case derives from the apparent equivalence or similarity among the resources detritus, diatoms and ostracods which is not utilized by any single consumer species.

These examples reflect two subtle conjugate interpretations of holes in communal complexes: either a hole represents the incorporation of species into the community with a heterogeneous niche as in web 22A, or it represents an exploited homogeneity in the underlying space {K} as in web 36 where resources that are similar in some way, or that appear to be tightly associated, are not utilized by any single species. Implicit here is the notion that resources in the environment are ordered, and that this ordering or association is perceived similarly by all species involved. This ordering of association among resources allows us to speak sensibly about an objective notion of homogeneity or heterogeneity in the niche.

For heuristics it is useful to view the underlying space as points distributed in some n-dimensional correlation space |K| where proximity among resource points reflects their closeness of relationship as perceived by consumers. This may be based on a combination of factors such as size, taxonomy, spatial covariance etc. In a rough sense, this picture is the conjugate of the standard multivariate statistical setting for analyzing communities where consumer species are viewed as points distributed in a correlation space whose axes represent the various ecological factors (e.g., resources) defining closeness of relationship among species. Here instead of focusing on consumers we look at closeness of relationship among resources as perceived by consumers. This is presumably at least partially determined by objective factors such as spatial location, etc. If species choose resources at random ignoring the natural ordering in |K| then heterogeneous niches would be created producing a situation similar to the one in our random null hypothesis where the community complexes were filled with holes. These heterogeneous niches may represent forbidden combinations. On the other hand, if niches are homogeneous and consist of resources that are close to each other in |K| then holes may be avoided.

On a finer scale, such homogeneity in a species niche implies the existence of a local ordering among resources. Consider, for example, an insectivorous lizard community where each species is morphologically constrained to consume prey items of a given size. Insofar as large prey are more similar to middle-sized prey items than to small ones, resources may be objectively ordered in |K| along a single size dimension from large to small. In general this ordering must be followed so that a species that consumes large and small insects will also be required to take in middle-sized ones. It is easy to see that a hole might appear if the local ordering is not followed (Figure 3.11). Therefore, if resources in nature are correlated in some way, and if the niches of the consumer species acknowledge this ordering, then heterogeneous combinations that lead to the formation of holes in $K_C(R;\lambda^{-1})$ may be avoided. Such homogeneity was a starting assumption in MacArthur's (1972) work on the economics of consumer choice. Therefore it is significant that the documented absence of holes gives justification for much of the later work on resource partitioning that use this assumption implicitly (MacArthur, 1972 and May, 1974). Rules for assembling communities and their food webs should be based on local orderings in the resources.

As pointed out in web 36 another requirement for avoiding holes in $K_C(R;\lambda^{-1})$ is that |K| should be packed sufficiently by consumers so that clusters of similar resources will be used by a single species. This is a situation where no local ordering exists. For example, if three resource points are close and equidistant in |K| there is no local ordering. This means if one resource is chosen there is no order or precedence for incorporating the other two resources into the

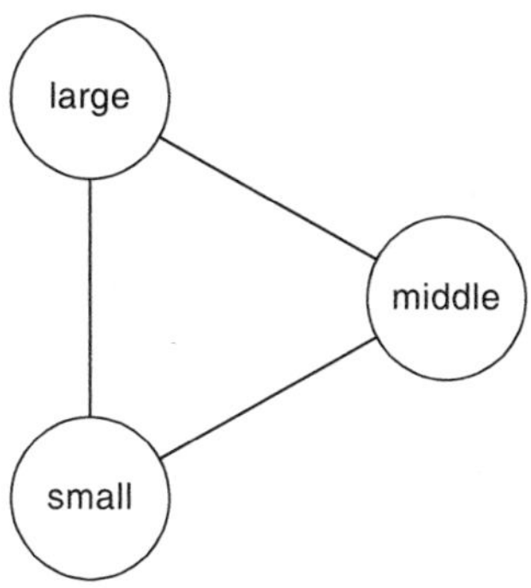

Figure 3.11 A hole may be created if a local ordering in the resource set is not followed.

niche. Equidistance or uniform density means that if all equivalent resources are not chosen by some consumer a hole may develop (cf. Figure 3.10).

Similarly holes may occur if resources in |K| are distributed in dense rings or with pockets (Figure 3.12). Therefore the absence of holes relies on the distribution of resource points in |K|, which depends ultimately on the local orderings perceived by consumers. In the next section we will see how it may be reasonable to view these points as distributed roughly with a convex density in |K| meaning that communities in nature may have tightly packed centers with less dense edges in |K|. Convex density helps to preempt holes in $K_C(R;\lambda^{-1})$.

3.3 ROBUSTNESS: DO HOLES APPEAR ACROSS VARIABLE THRESHOLDS?

So far our conclusion concerning the rarity of holes has been based on very coarse data from food web matrices where the interactions are recorded in simple binary form. A species either uses a resource or it does not. It is now of interest to ask how the homology of $K_C(R;\lambda^{-1})$ may be affected by using different thresholds to determine the presence or absence or linkages in $G_R(C;\lambda)$ and $K_C(R;\lambda^{-1})$. That is, is our find-

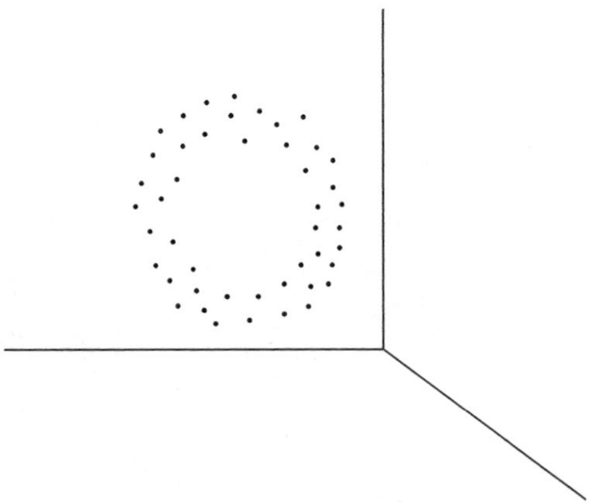

Figure 3.12 Holes may be created if the resources are distributed in rings or pockets in the correlation space |K|.

ing robust to the threshold chosen for deciding whether or not a link should be included?

3.3.1 Cody's Community Matrices

To do this we will examine a collection of 11 bird community matrices assembled by Cody (1974) from a diversity of new world habitats. The elements of these matrices α_{ij} have magnitudes related to the extent of niche overlap and were determined from a suite of behavioral and trophic characteristics. They are indicative of ecological similarity and presumably of the relative strength of competitive interaction. Because the community matrices here are symmetric ($\alpha_{ij} = \alpha_{ji}$) it will be straightforward to study the structure of the niche overlap graph $G_R(C;\lambda)$ at different thresholds, t, using the corresponding adjacency matrix A. The elements of A, a_{ij}, are constructed as follows: if $\alpha_{ij} \geq t$ then $a_{ij} = 1$, otherwise $a_{ij} = 0$. Setting $a_{ij} = 1$ in the adjacency matrix, places an edge between species i and j in $G_R(C;\lambda)$. Thus, beginning with a graph containing only the strongest linkages, by systematically lowering the threshold, t, we can study how the structure of $G_R(C;\lambda)$ changes with the addition of weaker links.

Although one cannot use the information in a community matrix to construct $K_C(R;\lambda^{-1})$ directly, it is possible to draw conclusions about its homology from properties of the corresponding niche overlap graph $G_R(C;\lambda)$.

3.3.1.1 Rigid Circuits and Holes

One useful characteristic of $G_R(C;\lambda)$ that relates to the homology of $K_C(R;\lambda^{-1})$ is the *rigid circuit property*. A graph G is said to be rigid circuit (Dirac 1965) if every circuitous path in G through more than three points has a chord (i.e., an edge which shortens the circuit). Such graphs have also been referred to as *chordal* (Harary 1976) and *triangulated* (Tarjan 1978). For example, the graph illustrated in Figure 3.13E is not a rigid circuit graph because there is a minimal circuit of length four. In Figure 3.13A, however, the non-rigid circuit (a, b, c, d) is cut by the chord (a, c) and there are no minimal circuits with three or more links. Thus, A is a rigid circuit graph. Further examples of this property are illustrated in Figure 3.13. Notice that graphs containing no circuits, that is *trees*, are rigid by default.

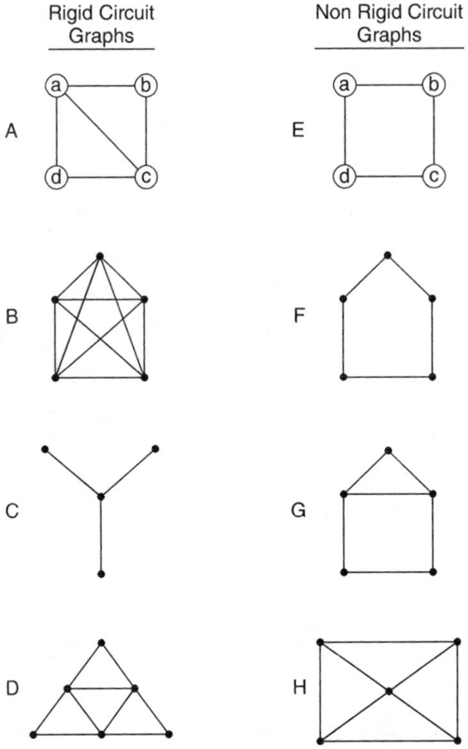

Figure 3.13 Examples illustrating the rigid circuit property. Rigid circuit graphs are also referred to as triangulated and chordal.

As will be discussed in section 4.4. under a mathematically deduced assembly rule, there is a clear analytic connection between rigidity in $G_R(C;\lambda)$ and the absence of holes in $K_C(R;\lambda^{-1})$. In this special case, rigidity in the niche overlap graph is necessarily associated with a lack of holes in the underlying complex and vice versa. Unfortunately, in the general case, there is no neat analytical result tying these two properties together. As seen in Figure 3.14 complexes lacking holes may have non-rigid overlap graphs, and rigidity in $G_R(C;\lambda)$ does not preclude holes. It is possible, however, to demonstrate that these two properties are related in a statistical sense. This is illustrated in Figure 3.15 where a tight statistical association is demonstrated between the absence of holes in $K_C(R;\lambda^{-1})$ and rigidity in $G_R(C;\lambda)$. These are random complexes generated to be free of holes by requiring that niches be homogeneous

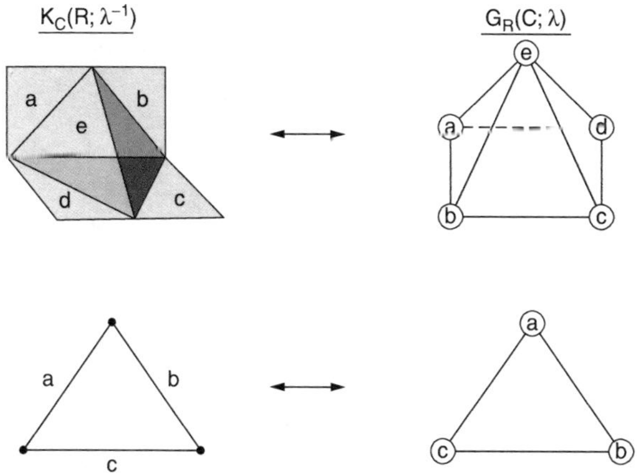

Figure 3.14 No necessary connection exists between rigidity in $G_R(C; \lambda)$ and the absence of holes in $K_C(R; \lambda^{-1})$.

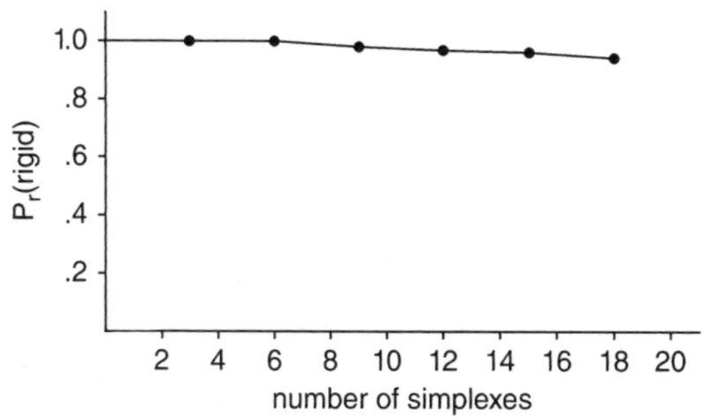

Figure 3.15 A tight statistical association exists between homological density in $K_C(R; \lambda^{-1})$ and rigidity in $G_R(C; \lambda)$. This graph shows that a complex constructed to be free of holes (cf. section 4. 3) has a very high probability of corresponding to a niche overlap graph that is triangulated. The random homologically dense complexes were constructed from a uniform mixture of 0, 1 and 2-simplexes.

(cf. section 4.4). Thus, for all practical purposes a community complex lacking holes will have a corresponding niche overlap graph which is triangulated. This point will be strengthened later on and the converse proved in section 4. 4.

One would predict therefore, that if the homology of $K_C(R;\lambda^{-1})$ discovered in the last section is robust, that $G_R(C;\lambda)$ will have the rigid circuit property at all thresholds of interaction strength. Using

Cody's data, this means that we should expect to find triangulated graphs at each step, as each edge corresponding to a lower threshold of α-value is added in the construction of $G_R(C;\lambda)$.

3.3.1.2 *Robustness of Rigid Circuits to Thresholds*

Figure 3.16A illustrates a representative sequence of niche overlap graphs at various thresholds of interaction strength for Cody's Wyoming Sagebrush community. In this case, at every level, that is for every edge added, the rigid circuit property was upheld. This same invariant topology was found in 4 out of 11 communities studied. Figure 3.16B and Appendix C present sequences of niche overlap graphs for the remaining 10 communities. In all, out of a total of 959 edges added (Table 3.2), only 27 resulted in non-rigid circuits and most of these became triangulated with the addition of the next few edges.[1] In fact, if a small measurement error of ± 0.05 is allowed in Cody's overlap estimates (α_{ij}) then one could argue strongly that all 959 edges could be added in a manner that preserves rigidity in $G_R(C;\lambda)$.

Thus, these limited results on bird communities suggest that the absence of holes is independent of the threshold chosen and as such

Table 3.2 Cody's (1974) Bird Community Matrices

	Community	Species	Edges Added	Non-Rigid Edges
1	Wyoming Willows	12	66	0
2	Wyoming Sagebrush	5	10	0
3	Colorado Sagebrush	5	10	0
4	Mohave Desert	12	66	3
5	California Chapparal	15	105	2
6	Sonoran Desert	16	120	4
7	Arizona Mesquite	14	91	1
8	Arizona Oak-Pine	20	190	5
9	Coastal Scrub, Chile	5	10	0
10	Mattoral, Chile	19	171	8
11	Savanah, Chile	16	120	4

[1]As a first indication of the significance of this result the average number of non-rigid edges expected for a random graph with 15 vertices (c.f. case 5 above) is 23.6. Full null results will be deferred to a later date.

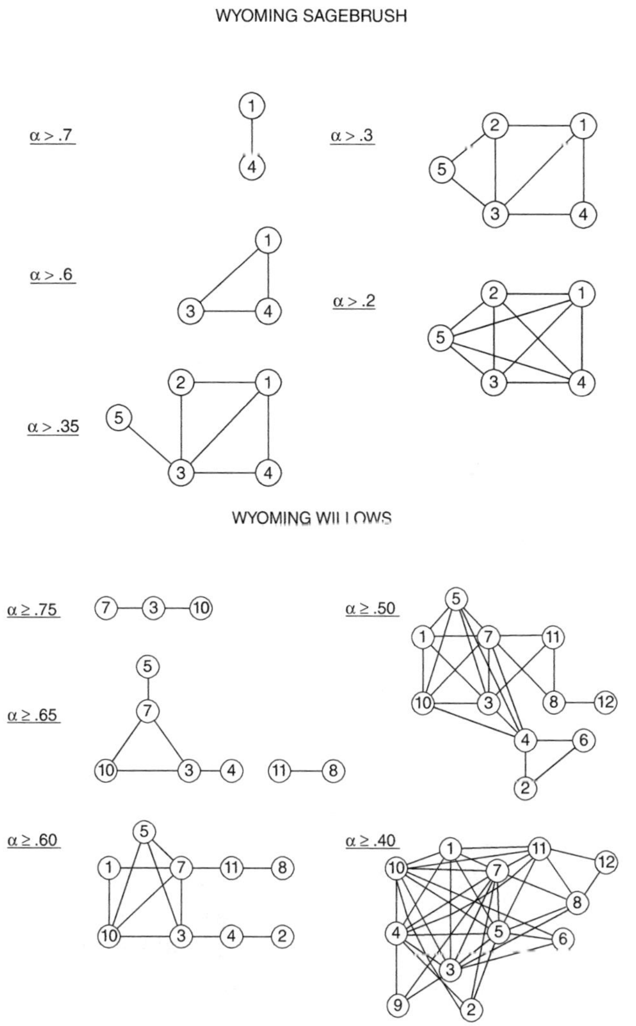

Figure 3.16 Real niche overlap graphs at various thresholds, $\alpha_{ij} \geq t$, tend be to triangulated.

a) Cody's (1974) Wyoming Sagebrush community
1) Brewer's sparrow
2) Cliff swallow
3) Vesper sparrow
4) W.C. sparrow
5) Brewer's blackbird

b) Cody's (1974) Wyoming Willows community
1) C. hummingbird
2) Wilson's warbler
3) Yellowthroat
4) Yellow warbler

is a robust finding. It should be mentioned that rigidity in $G_R(C;\lambda)$ places a limit on the size of the holes that could possibly occur in $K_C(R;\lambda^{-1})$ by forbidding non-bounding cycles larger than 3. That is, in the rare case when rigidity allows a hole, that hole can be no larger than a triangle.

3.3.2 Quasiconvexity in the Niche

We now examine how homological robustness implies that community niches (*sensu* Hutchinson-Whittaker) have densely packed centers and less well occupied peripheries (Whittaker, 1977 and Ricklefs and Travis, 1980).

Consider a 2-dimensional resource space where each species has a convex utilization function as shown in Figure 3.17. The union of these overlapping utilization functions for all species forms a landscape of hills and valleys comprising a so-called *utilization surface* for the community. In general, given equal symmetrical utilization functions, species whose niches are joined by higher ridges overlap more extensively and have higher α-values than those joined by lower ridges. One can study the interconnections among species at different thresholds by examining different *level curves* of the landscape parallel to the resource plane (Figure 3.17). Higher level curves cut the mountain tops and high ridges, portraying the strong interactions; lower cross sections fill out the picture by including the weaker links between species. In each slice, species niches may be viewed as overlapping elliptical discs. Figure 3.17 demonstrates how the method of trophic equivalence (Chapter 2) may be applied to construct the corresponding complex and niche overlap graph. Clearly these cross sectional slices and $K_C(R;\lambda^{-1})$ are homological duals in that at a given threshold a hole in a complex corresponds to a hole in the thin section and vice versa. By stacking these slices in order, we can see that the presence of holes at a given threshold corresponds to utilization surfaces having circuitous ridges and closed basins. On the other hand, communities possessing robust homologies with no holes at any level will have *quasi-convex* utilization surfaces capable of shedding water without forming pools. A quasi-convex surface is one where every point has a directional derivative which is less than or equal to zero in some direction. Here the ridges and valleys rather than forming

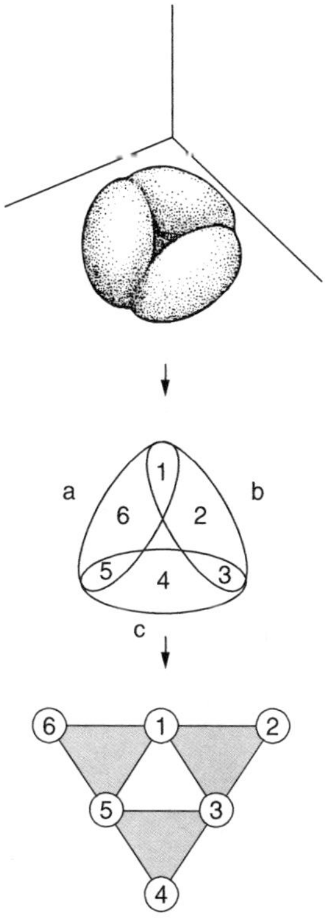

Figure 3.17 The relationship between the geometry of a utilization surface and holes in $K_C(R; \lambda^{-1})$.

A) A utilization surface for a group of three species having a central crater.
B) A thin-section of the utilization surface at a given level of interaction strength. Numbers indicate different resources defined according to trophic equivalence.
C) The corresponding complex $K_C(R; \lambda^{-1})$.

This figure shows how a crater in the utilization surface corresponds to a hole in $K_C(R; \lambda^{-1})$ at some threshold level.

closed chains and pools will tend to cascade smoothly from higher ridges near the center of the landscape to lower ones near the edges.

Therefore, Cody's data suggest that utilization surfaces are quasiconvex with species packed tightly along high central ridges and less tightly in the periphery (cf. examples in Appendix C).

This meshes with an alternative and interesting analysis of these same communities by Ricklefs and Travis (1980) who applied a principal components analysis that used *morphological* characteristics in place of behavioral and trophic ones (see also Gatz (1979)). They found that all of Cody's communities possessed a core of species occupying the first 3 to 4 principal components, with additional species attached peripherally bringing into play other minor axes. If one relates morphology to resources use (Gatz, 1979), these core species define a set of core resources over which species are tightly packed. Insofar as species tend to pack densely over the set of core resources and less densely in the periphery, the utilization surface will have high central ridges with lower ones in the periphery. This image resonates with the quasi-convex picture of the niche uncovered above, using a different analysis on the same communities.

A quasi-convex utilization surface means that $|K|$ should have a convex density which as already suggested in section 3.2.3 dovetails with the absence of holes in $K_C(R;\lambda^{-1})$.

SUMMARY

The problem of minimal packing in communities was addressed by studying the homological properties of the simplicial-complex representation of the niche. The aim was to determine whether niche space is conservatively packed in a manner that precludes holes or whether it is loosely assembled to look more like a multidimensional swiss cheese. Results from food web data on 102 real (73 nontrivial) communities show that all webs, with the possible exception of two cases, are homologically dense and conservatively packed. This conclusion was strengthened by eleven detailed studies on bird communities, where the absence of holes appeared to be independent of the thresholds chosen for determining the existence of interactions. The interpretation given to this result was that the utilization surfaces for these communities may be quasi-convex; that is, capable of shedding water. Interestingly, this interpretation is shown to coincide with an alternative independent analysis by Ricklefs and Travis (1980), which suggest that these same communities may have a densely packed core of similar species with a weaker periphery of more novel ones. In the next chapter, we will see how this conservative structure in the niche may be explained in terms of a very simple assembly rule for community foodwebs.

4

INTERVALITY: TIP OF AN ICEBERG

The last chapter examined expanded views of the niche provided by $K_C(R;\lambda^{-1})$ and $G_C(R;\lambda^{-1})$. Here we focus on regularities observed in the condensed picture of niche overlaps contained in $G_R(C;\lambda)$, the consumer overlap graph.

A recent discovery by Joel Cohen (1978) is the broadly based empirical finding that the diet overlaps in $G_R(C;\lambda)$ for most real communities, can be collapsed down into a simplified 1-dimensional representation known as an *interval graph.* As illustrated in Figure 4.1A, ensembles with this property can be reduced to a linear sequence of overlapping niches. Loosely speaking, such systems may be thought of as being organized with species ordered along a single composite niche axis. The fact that real niche spaces are contained within this narrow subset of mathematically possible topologies is more than a curiosity. It is a fundamental result about their structure, and is important in that it attempts to give support to other areas of ecological theory such as the species packing problem, that are best understood in the 1-dimensional case (May 1974). That intervality may be an expression of deeper constraints acting on the assembly of communities is provocative and may help illuminate how large multispecies communities can persist when their randomly constructed model analogues are so dynamically fragile.

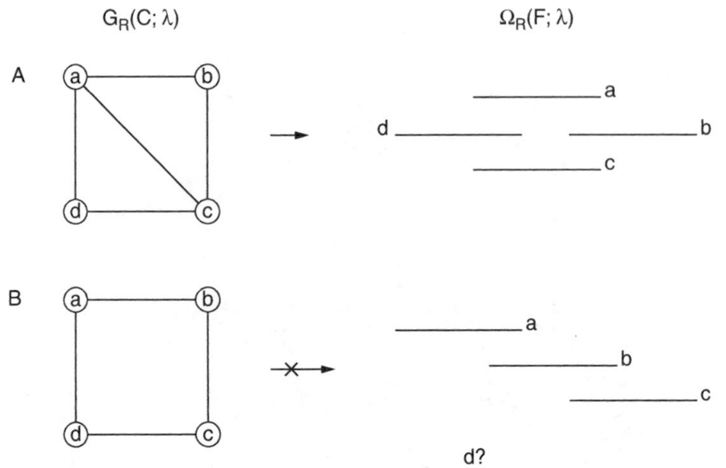

Figure 4.1 An illustration of intervality: the niche overlap graph A can be reduced to a set of overlapping intervals on the line whereas B cannot.

The curious fact that most real food webs are interval was left unexplained by Cohen. He constructed seven plausible null hypotheses and demonstrated that none of them could account for the observed excess of interval webs. It will be my aim here to demonstrate that this simple topology in $G_R(C;\lambda)$ can be understood in terms of deeper constraints on the homology of $K_C(R;\lambda^{-1})$ that were observed in Chapter 3, and that these regularities, in turn, can be explained by a mathematically deduced and biologically reasonable community assembly rule.

4.1 NECESSARY AND SUFFICIENT CONDITIONS FOR INTERVALITY

A niche overlap graph $G_R(C;\lambda)$ is said to be *interval* if it can be collapsed into a set of overlapping intervals on the line, that is, if it is isomorphic to some graph $\Omega_R(F;\lambda)$ where F is a family of intervals. Here it is understood that intervals in $\Omega_R(F;\lambda)$ that correspond to consumer species or vertices in $G_R(C;\lambda)$ are allowed to overlap if and only if their corresponding vertices are adjacent (joined) in G.

$$v_i \text{ adj } v_j \text{ in } G \Leftrightarrow f_i \quad f_j \neq \phi \text{ in } \Omega$$

Thus two species i and j that overlap with respect to some resource is represented in Ω as two overlapping intervals. Figure 4.1 illustrates an example of a niche overlap graph that satisfies this condition (4.1A) and one that does not (4.1B). Notice that in 4.1B species d cannot be added to Ω as a single interval, thus 4.1B is not an interval graph. Intervality is a special case of *boxicity* (Roberts, 1969) which is the minimum number of dimensions required of boxes to represent adjacency in G. Intervals are simply 1-dimensional boxes, and an interval graph therefore has boxicity 1.

A useful set of necessary and sufficient conditions for characterizing interval graphs derives from the following two properties.

1. A graph G is said to be *rigid circuit* (Dirac, 1965) if every circuitous path in G through more than three points has a chord. This property has already been introduced in section 3.3 (see Figure 3.13).
2. A graph is *asteroidal* (see Figure 4.2A) if it contains three distinct points v_0, v_1, v_2 and three paths W_0, W_1, W_2 such that for $i = 0, 1, 2$
   (i) if the path W_j connects two points v_j and v_k (j, k, ≠ i)
   (ii) then v_i is not a neighbor of W_j (i.e., v_i is more than one edge away from W_j)

Hence, each point v_i of this triple is not a neighbor of a path connecting the other two points. Such a triple of points v_0, v_1, v_2 is called an *asteroidal triple* (Lekerkerker and Boland 1962). In Figure 4.2A, for example v_0 is not a neighbor of W_j and likewise v_1 and v_2 are not neighbors of W_2 and W_0 respectively. This tree graph, therefore, is asteroidal and v_0, v_1, v_2 is an asteroidal triple. On the other hand, in Figure 4.2B, v_2 is indeed a neighbor of the path W_0; hence 4.2B is not asteroidal. The reader may wish to confirm that the only graph in Figure 4.3 that is asteroidal is 4.3C.

We may now characterize an interval graph as satisfying the following two conditions (Lekerkerker and Boland 1962). A graph G is interval if and only if:

1. G is a rigid circuit graph
2. G is not asteroidal (4.1)

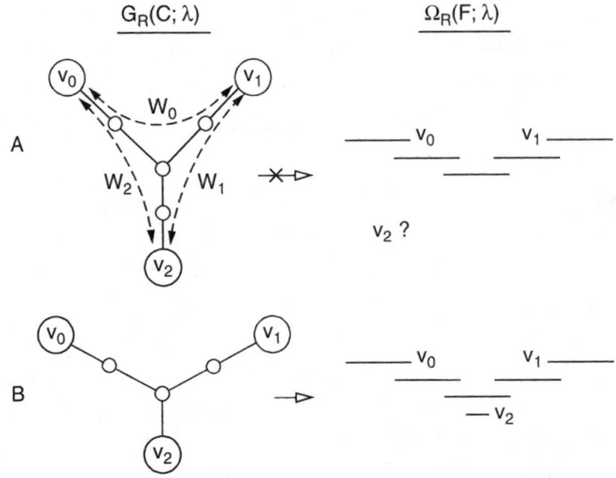

Figure 4.2 An illustration of the asteroidal property: Graph A is asteroidal (v_0, v_1, v_2 is an asteroidal triple) whereas B is not.

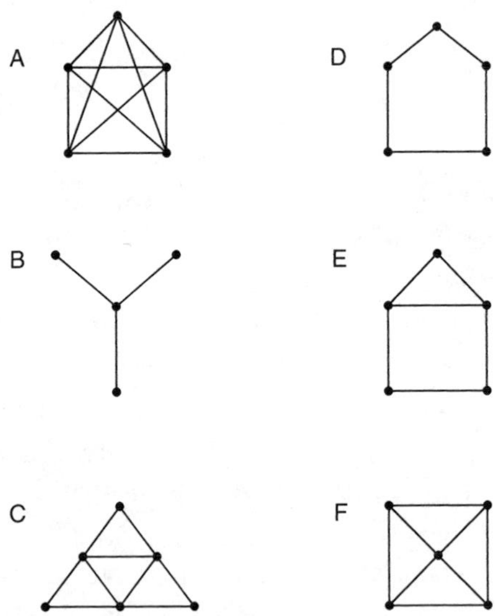

Figure 4.3 Examples of rigid circuit and asteroidal graphs: Graphs D, E, F are not rigid and C is asteroidal. The remainder are rigid and non-asteroidal (therefore interval by (4.1)).

Thus the graphs D, E and F in Figure 4.3 are not interval because they fail to satisfy the first condition, whereas the triangulated graph in Figure 4.3C and the tree graph in Figure 4.2A are not interval because they are asteroidal. The graphs in 4.3A, B and 4.2B, on the other hand, satisfy both conditions and are interval. Therefore, real niche overlap graphs $G_R(C;\lambda)$ that are interval must have the rigid circuit property and must also not be asteroidal.

4.2 RIGID CIRCUITS: A ROBUST PROPERTY OF REAL COMMUNITIES

To understand why niche overlap graphs $G_R(C;\lambda)$ for most real communities are interval we ask which of the above conditions (4.1) fails to be satisfied in those communities that cannot be reduced to interval graphs.

Table 4.1 presents results for the 73 nontrivial community webs from Briand's collection (Table 3.1), indicating those that are rigid, not asteroidal and interval. In total, only 10 communities have niche overlap graphs that are not interval. Of these 10, it is interesting that all fail to satisfy the second condition of (4.1): that is, all non-interval graphs are asteroidal. Moreover, the first condition, the rigid circuit property, appears significant as it is met in every community niche overlap graph studied here. One may wonder, therefore, whether rigidity in itself is sufficient to account for the observed high frequency of interval graphs, that is, whether rigidity is the more fundamental property—so that, the prevalence of intervality (the rarity of asteroidal graphs) may arise as a simple by-product of rigidity.

Figure 4.4 shows that the rigid circuit property is not a general characteristic of random connected graphs. Most real communities have connectivities (observed number of edges/maximum possible number of edges) in the middle range, between 0.4 to 0.7, where the probability of generating rigid circuit graphs at random is low. On the other hand, as seen in Chapter 3, real communities are not random structures. In fact, the apparent ubiquity of triangulated niche overlap graphs may be predicted from the fact that the corresponding complexes $K_C(R;\lambda^{-1})$ for these communities were homologically dense. This brings the argument a step further back, suggesting the

Table 4.1 Statistics on $G_R(C;\lambda)$ Calculated for 73 Real Communities

Web I.D.	p	q	Rigid Circuit	Asteroidal	Interval
1A	4	6	1	0	1
B	3	3	1	0	1
2A	7	15	1	0	1
B	4	3	1	0	1
3A	10	18	1	1	0
B	7	8	1	1	0
C	2	1	1	0	1
4A	10	19	1	0	1
5A	3	3	1	0	1
B	2	1	1	0	1
6A	8	24	1	0	1
B	13	27	1	1	0
C	2	1	1	0	1
7A	12	28	1	0	0
B	3	3	1	0	1
8A	11	26	1	0	1
9A	5	10	1	0	1
B	2	1	1	0	1
10A	2	1	1	0	1
11A	2	1	1	0	1
B	2	1	1	0	1
12A	4	6	1	0	1
B	3	3	1	0	1
C	3	3	1	0	1
13A	4	6	1	0	1
B	4	6	1	0	1
C	2	1	1	0	1
14A	5	7	1	0	1
15A	5	10	1	0	1
16A	5	8	1	0	1
B	4	5	1	0	1
17A	10	11	1	0	1
18A	15	34	1	1	0
B	3	3	1	0	1

Web I.D.	p	q	Rigid Circuit	Asteroidal	Interval
19A	12	30	1	0	1
B	2	1	1	0	1
20A	14	29	1	0	1
B	2	1	1	0	1
21A	7	17	1	0	1
22A	21	72	1	1	0
B	4	6	1	0	1
23A	14	44	1	0	1
24A	5	7	1	0	1
B	3	3	1	0	1
25A	6	9	1	0	1
B	11	43	1	0	1
C	3	3	1	0	1
26A	24	64	1	1	0
27A	15	43	1	1	0
B	3	3	1	0	1
28A	9	36	1	0	1
B	8	15	1	0	1
C	13	31	1	0	1
D	3	3	1	0	1
E	2	1	1	0	1
29A	4	4	1	0	1
C	7	21	1	0	1
D	7	18	1	0	1
30A	11	31	1	0	1
31A	12	50	1	0	1
32A	12	50	1	0	1
33A	25	164	1	1	0
B	2	1	1	0	1
34A	8	24	1	0	1
35A	11	45	1	0	1
36A	14	64	1	0	1
37A	12	39	1	0	1
B	2	1	1	0	1
C	4	6	1	0	1

Web I.D.	p	q	Rigid Circuit	Asteroidal	Interval
38A	28	258	1	0	1
39A	34	189	1	1	0
B	2	1	1	0	1
40A	8	18	1	0	1

Food webs from Briand, 1982.
p = number of species (vertices in $G_R(C;\lambda)$).
q = number of overlaps (edges in $G_R(C;\lambda)$).

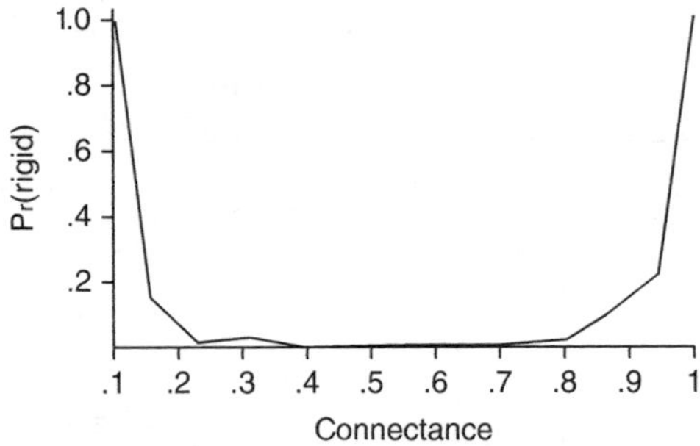

Figure 4.4 The rigid circuit property is not a general characteristic of random connected graphs. For these random graphs p = 20. Pr(rigid) = the probability that a random connected graph is rigid as estimated from 1000 replicates. Random connected graphs were not simple Erdos-Renyi constructions, but were constructed by first producing a random spanning tree and then adding edges between vertex pairs chosen with equal probability.

possibility that the absence of holes in $K_C(R;\lambda^{-1})$ may suffice to account for both rigidity and intervality in $G_R(C;\lambda)$.

4.3 HOLES, RIGIDITY, AND INTERVALITY

As pointed out in Chapter 3 the relationship between the rigid circuit property in $G_R(C;\lambda)$ and a lack of holes in $K_C(R;\lambda^{-1})$ is not generic but emerges as a tight statistical association within the realm of natural systems. We can now determine exactly how tight this association is for our sample of real communities and ask whether having a trivial homology (a complex with no holes) in $K_C(R;\lambda^{-1})$ is sufficient to account for the observed number of triangulated and interval niche overlap graphs.

To answer this question, for each community in Table 4.1, we construct a sample of random community complexes lacking holes. These are then tested to determine the frequency of rigid and interval niche overlap graphs. Recall from Chapter 3 that a trivial homology requires that niches be combined in $K_C(R;\lambda^{-1})$ in a conservative fashion. That is, each species clique or simplex must be added to the community in a way that insures homogeneity in the niche. If in the sequential construction of a community each additional species joins the growing complex K_p (*parent complex*) without forming a hole, then the final community complex will also be free of holes. Specifically, holes will not be formed in $K_C(R;\lambda^{-1})$ if the set of resources (*s-cutset*) by which each additional species joins to the parent complex K_p is sufficiently homogeneous—i.e., if the resources in the s-cutsets are sufficiently closely related. These s-cutsets can be made sufficiently homogeneous to preclude the formation of holes by requiring that the resources in each s-cutset be adjacent to some *seed* or *characteristic* resource v_0 in the parent complex K_P. That is, v_j is an element of the s-cutset *only if* v_j adj v_0 or $v_j = v_0$ in K_P. v_0 then forms the nucleus of the s-cutset. It is easy to see that this condition is sufficient to guarantee a trivial homology since choosing resources that are adjacent to v_0 eliminates the possibility of creating non-bounding cycles in $K_C(R;\lambda^{-1})$. For the moment we will take this as a sufficient condition for avoiding holes. In section 4.5 we will see that it can be derived as a necessary condition from the structure of $G_R(C;\lambda)$.

Using the observed clique vector (Table 3.1) for each community, corresponding random complexes may be constructed by choosing the order of incorporation of species, the successive seed resources v_0, and the size of the s-cutsets with uniform random distributions over appropriate limits. Essentially, this is the same procedure implemented in section 3.2.2 with the added requirement, v_j ε s-cutset only if:

$$v_j \text{ adj } v_0 \text{ in } K_P.$$

For each community in Table 4.1, 1000 random homologically dense complexes were generated as above.[1] Results on the estimated probabilities of obtaining rigid and interval graphs are presented in section A of Table 4.2. Although the figures here are for species simplexes chosen in random order, virtually identical results are generated when simplexes are chosen in order of size from largest (generalist) to smallest (specialist).

Looking at rigidity first, summing the expected frequencies for all communities we find that 72.07 of these have triangulated niche overlap graphs. This appears to agree with the 73 rigid circuit graphs that were observed, and the difference is not significant, $z = 1.04$. Therefore, conservative packing is sufficient to explain the tendency for niche overlap graphs to be rigid.

Column 2 of section A presents the predicted frequency of interval graphs generated by this procedure. Again a very good fit is realized; 64.15 are predicted and 63 are observed, $z = 0.079$.

Figure 4.5 presents the distribution of *P* values obtained by comparing the observed overlap in $G_R(C;\lambda)$ (i.e., number of edges, q) to the overlap that is generated by random conservative packing in $K_C(R;\lambda^{-1})$ (cf. Cohen 1978). As discussed in section 3. 3, if we are correct here in predicting the number of edges q in $G_R(C;\lambda)$ we should expect the values of *P* to be distributed uniformly in the interval 0 to 1. As can be seen, our minimal packing procedure stands in reasonable agreement on this score also. Thus a conservative homology in $K_C(R;\lambda^{-1})$ appears to be sufficient to explain the ubiquity of triangulated and interval niche overlap graphs observed in nature.

Indeed, conservative packing can also be shown to be necessary since these results tend to be very sensitive as to whether or not holes are allowed in the construction of $K_C(R;\lambda^{-1})$. In particular, if

[1]Efficient algorithms for implementing this procedure and for detecting rigid circuit and interval graphs were derived using the C programming language. A zero fill-in ordering was used to determine rigidity (Booth and Leuker, 1976), and the criterion of consecutive 1's in the dominant clique matrix for $G_R(C;\lambda)$ was applied to detect interval graphs (Fulkerson and Gross 1965; also see Cohen 1978). For the relatively small size of dominant clique matrices encountered in Table 4.1 the polynomial time procedure used here for testing consecutive 1's was considerably faster than the linear time P-Q tree algorithm suggested by Booth and Leuker (1976); they obtain an $O(m + n + f)$ complexity bound for $m \times n$ matrices having f non-zero elements.

Table 4.2 Statistics on $G_R(C;\lambda)$ for 73 Real Communities

	A NO HOLES, d = 1		B SMALL HOLES, d ≤ 2		C ASSEMBLY
Web I.D.	Rigid Circuit	Interval	Rigid Circuit	Interval	Perfect Addition
1A	1	1	1	1	1
B	1	1	1	1	1
2A	1	1	1	0.99	1
B	1	1	1	1	1
3A	1	1	1	0.97	0.63
B	1	1	0.89	0.88	0.89
C	1	1	1	1	1
4A	0.99	0.91	0.92	0.77	0.67
5A	1	1	1	1	1
B	1	1	1	1	1
6A	1	0.97	1	0.92	1
B	1	0.61	0.81	0.36	0.30
C	1	1	1	1	1
7A	0.98	0.70	0.82	0.46	0.56
B	1	1	1	1	1
8A	1	0.81	0.89	0.60	0.69
9A	1	1	1	1	1
B	1	1	1	1	1
10A	1	1	1	1	1
11A	1	1	1	1	1
B	1	1	1	1	1
12A	1	1	1	1	1
B	1	1	1	1	1
C	1	1	1	1	1
13A	1	1	1	1	1
B	1	1	1	1	1
C	1	1	1	1	1
14A	1	1	1	1	1
15A	1	1	1	1	1
16A	1	1	1	1	1
B	1	1	1	1	1
17A	1	0.95	1	0.94	0.52

	A NO HOLES, d = 1		B SMALL HOLES, d ≤ 2		C ASSEMBLY
Web I.D.	Rigid Circuit	Interval	Rigid Circuit	Interval	Perfect Addition
18A	1	0.90	1	0.76	0.19
B	1	1	1	1	1
19A	1	0.74	0.89	0.59	0.59
B	1	1	1	1	1
20A	0.99	0.54	0.85	0.34	0.21
B	1	1	1	1	1
21A	1	0.99	0.86	0.82	1
22A	1	0.94	0.11	0.26	0
B	1	1	1	1	1
23A	1	0.54	0.77	0.24	0.57
24A	1	1	1	1	1
B	1	1	1	1	1
25A	1	1	1	1	1
B	1	0.95	1	0.87	0.99
C	1	1	1	1	1
26A	0.86	0.08	0.25	0	0.01
27A	0.95	0.45	0.73	0.16	0.37
B	1	1	1	1	1
28A	1	1	1	1	1
B	1	0.99	1	1	0.93
C	1	0.99	1	0.93	0.44
D	1	1	1	1	1
E	1	1	1	1	1
29A	1	1	1	1	1
C	1	1	1	0.99	1
D	1	0.99	1	0.97	1
30A	0.98	0.74	0.77	0.49	0.82
31A	0.95	0.65	0.48	0.18	0.99
32A	0.93	0.51	0.53	0.20	0.99
33A	0.97	0.22	0.52	0.03	0.33
B	1	1	1	1	1
34A	1	0.95	0.79	0.69	1
35A	0.97	0.73	0.54	0.33	1

	A NO HOLES, d = 1		B SMALL HOLES, d ≤ 2		C ASSEMBLY
Web I.D.	Rigid Circuit	Interval	Rigid Circuit	Interval	Perfect Addition
36A	0.98	0.42	0.56	0.19	1
37A	0.98	0.74	0.87	0.44	0.84
B	1	1	1	1	1
C	1	1	1	1	1
38A	0.76	0.01	0.03	0	0.70
39A	0.82	0.02	0.23	0.01	0.01
B	1	1	1	1	1
40A	1	0.94	0.98	0.85	0.98
TOTALS	72.06	63.15	65.24	57.97	62.36
	z = 1.04	z = 0.08	z = 4.1	z = 2.5	z = 0.33
	From Table 4.1:	Observed rc = 73 Observed int = 63			

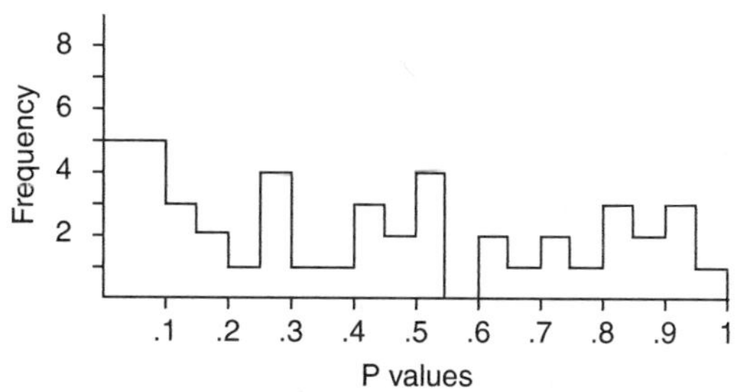

Figure 4.5 The distribution of P values for the observed overlap versus the number of niche overlaps predicted by conservative packing.

the packing procedure is modified slightly so that resources up to a distance of 2 from v_0 are allowed in the s-cutsets, then small three-sided holes are possible. As shown in section B of Table 4.2 this minor modification renders the procedure incapable of accounting for the observed frequencies of either rigid circuit or interval graphs. The respective z values here are z = 4.09 and z = 2.50, both indicating strong significant differences.

Roughly speaking, conservative packing in $K_C(R;\lambda^{-1})$ appears to be both necessary and sufficient to explain the ubiquity of rigid circuit graphs and the frequency of interval niche overlap graphs in Table 4.1; no holes was sufficient, small holes was not. This means that the regularities in $G_R(C;\lambda)$ may be understood in terms of deeper constraints acting on the homology of $K_C(R;\lambda^{-1})$.

It may be justifiable to point the causal arrow from homology to intervality or from rigidity to intervality. However, it is not correct to view the lack of holes in $K_C(R;\lambda^{-1})$ as the *cause* of triangulated overlap graphs. Rather, as will be demonstrated in the next section, both the absence of holes in $K_C(R;\lambda^{-1})$ and rigidity in $G_R(C;\lambda)$ are alternative sides of the same coin. Both properties can be explained as manifestations of a common rule for assembling communities. Curiously, this rule does not reveal itself initially through the more detailed representation $K_C(R;\lambda^{-1})$, but instead becomes first apparent in the condensed portrait of $G_R(C;\lambda)$.

4.4 ASSEMBLY RULES IMPLIED BY THE RIGID CIRCUIT PROPERTY IN $G_R(C;\lambda)$

A central problem in ecology involves finding and understanding constraints acting on the construction of ecological ensembles. This problem has been manifested in studies of energetics (Lindeman, 1943), succession (Curtis, 1953; Horn, 1971), colonization (Patrick, 1968), and in interpreting biogeographic patterns in the data (Diamond, 1976; Simberloff, 1978; Heatwole and Levins, 1978). My aim here will be to demonstrate that the rigid circuit property in $G_R(C;\lambda)$ implies the operation of specific biological constraints in the assembly of real communities. We shall see how the final structure of the niche overlap graph can reveal key information about its process of generation.

Root (1970) has introduced the term *guild* to denote a tightly-knit group of functionally related species. Thus one can speak of Paine's rocky intertidal community as a guild or MacArthur's warblers as a guild. Associated with this term is an obvious graphical analogue in the term "clique" as it is realized in the consumer overlap graph $G_R(C;\lambda)$. Recall from section 2.1 that clique is taken to mean a complete (maximally connected) subgraph. Therefore, a clique in $G_R(C;\lambda)$ represents a group of species whose niches mutually overlap (although not necessarily with respect to the same resource). In the coloquial sense, guilds have their most natural correspondence with the notion of *maximal* or *dominant clique*, that is, a clique contained in no larger clique. It may be useful however to regard the term guild in a more flexible and realistic manner so that it can be applied to different degrees of functional relatedness. In this sense a guild may correspond to a maximal clique or to any clique contained within a maximal clique. For the moment, the operational requirement for the usage of "guild" here is that species be mutually connected in $G_R(C;\lambda)$.

The assembly rule that I will derive and that is implied by the rigid circuit property is that communities tend to be built up in a conservative manner that involves adding species to single guilds (cliques) rather than by bridging multiple guilds. That is, new species as they come into the community will tend to compete exclusively within a single existing guild. These single guilds may be termed *relatively minimal c-cutsets* to denote the smallest clique whose removal isolates the added vertex i.e., the clique to which the new species is tethered. For example, in Figure 4.6A species d is added to the single guild defined by species a and c, that represents a relatively minimal c-cutset. This sequence I will demonstrate is allowed in natural systems whereas the sequence in Figure 4.6B is generally forbidden. Here species c is not added to the community through a relatively minimal c-cutset, rather it is added non-conservatively by bridging two guilds, (a,d) and (a,b). Therefore, I shall attempt to show that real communities proliferate by the sequential incorporation of new species into single guilds (relatively minimal c-cutsets) as in Figure 4.6A, rather than across multiple non-nested guilds as in Figure 4.6B.

The only additional requirement here is that species be added to communities in order of increasing (or equal) specialization, from generalist to specialist. This is a reasonable constraint, I believe, and is a trend that has been widely observed and is regarded as conventional wisdom (e.g., Krebs, 1973; Whittaker, 1972). The degree of specialization is taken here to be inversely related to the point degree of a species represented as a vertex in $G_R(C;\lambda)$. *Point degree* denoted deg(v) (Harary, 1969) is the number of edges radiating from a vertex. A species therefore that overlaps with many others (high point degree) is more of a generalist than one that overlaps with a few.

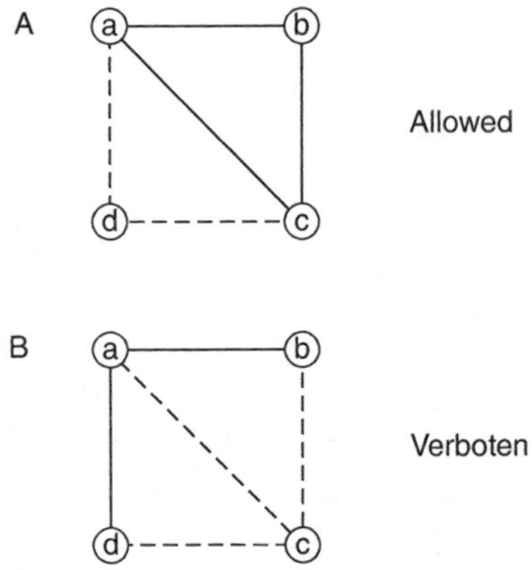

Figure 4.6 Two alternative incorporation sequences in $G_R(C;\lambda)$: In A the last species d is added to a single guild (clique), in B species c is added by bridging two guilds, (a, d) and (a, b).

We can now prove that if communities are rigid and are assembled in order of increasing specialization (decreasing point degree as determined from the final overlap graph $G_R(C;\lambda)$) that species must be added to the complex one guild at a time rather than by bridging multiple guilds.

Dirac (1965) has shown that every rigid circuit graph possesses at least one *extreme vertex*, or a vertex whose neighbors form a clique. For example, in Figure 4.7 points 1, 2, and 3 are extreme vertices whereas 4, 5, and 6 are not. Notice that if one eliminates an extreme vertex the remaining graph will continue to be rigid. In fact one can completely collapse a rigid circuit graph by sequentially eliminating extreme vertices. Such an ordering on the vertex set V is known as *perfect elimination ordering*. Not only do all rigid circuit graphs have a perfect elimination ordering but conversely any graph that has this property must be a rigid circuit graph (Rose et. al. 1976, Booth and Leuker, 1976).

Perfect elimination graphs are also characterized as possessing a *zero fill-in order*. If α is a numbering on the vertex set V of a graph

$$\alpha: V \rightarrow \{1, 2, 3, \ldots, n\}$$

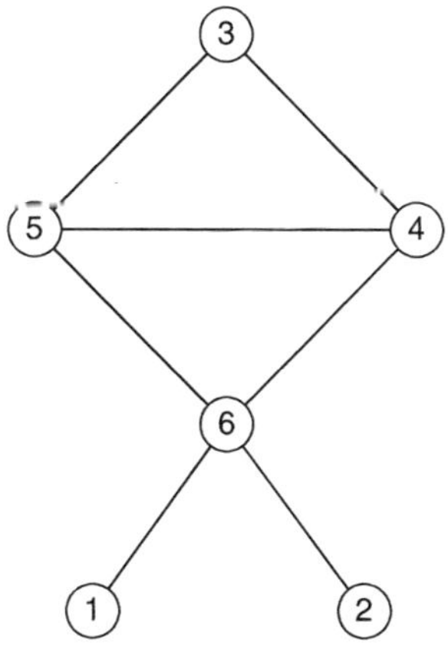

Figure 4.7 A perfect elimination ordering: Points 1, 2 and 3 are extreme vertices, whereas 4, 5 and 6 are not.

then the *fill-in* produced by ordering this set is:

$$F(\alpha) = \{(v_0, v_1) | (v_0, v_1) \notin E \text{ and } \exists \text{ a path from } v_0 \text{ to } v_1 \text{ containing only } v_0 \text{ and } v_1 \text{ and vertices numbered less than } \min[\alpha(v_0), \alpha(v_1)]\}$$

where E is the edge set. For example in 4.6B the fill-in would consist of the edge (b.d). A perfect elimination graph is defined as having a zero fill-in ordering, i.e., $F(\alpha) = \emptyset$ (Rose et. al. 1976).

Therefore real niche spaces $G_R(C;\lambda)$ that are rigid can always be disassembled by eliminating species that are attached to single guilds (cliques). That is, they can be collapsed by sequentially eliminating extreme specialists. Conversely, it can be said that such niche overlap graphs possess a *perfect addition ordering* in that they can always be constructed by sequentially attaching species to individual guilds (relatively minimal c-cutsets).

To say that such an ordering *exists*, however, is far from saying that a given community was actually assembled in this manner. It is at this stage that the assumed sequence from generalist to specialist becomes important. We wish to prove that such a sequence necessarily implies a perfect addition ordering (adding to just one clique). It will suffice here to show that by reversing the sequence and eliminating specialists first, one has a zero fill-in order. For this we require a property of orderings that is strong enough to satisfy zero fill-in on rigid circuit graphs and that may also possibly be met by the ordering implied by the point-degree of a graph. Following Tarjan (1978) such a property is:

If $\alpha(v_0) < \alpha(v_1) < \alpha(v_2)$ and $(v_0, v_2) \varepsilon E$ but $(v_1, v_2) \not\varepsilon E$, then $\exists$ a vertex v^* such that $\alpha(v_1) < \alpha(v^*)$ and $(v_1, v^*) \varepsilon E$ but $(v_0, v^*) \not\varepsilon E$

Because any ordering α of a rigid circuit graph that satisfies (4.2) is a perfect elimination ordering (Tarjan, 1978), all that remains is to show that ordering vertices in G by their point degree also satisfies (4.2).

Let α be an ordering on the vertex set V such that $\alpha(v_i) < \alpha(v_j)$ if and only if $\deg(v_i) \leq \deg(v_j)$. Following (4.2) suppose that $\alpha(v_0) < \alpha(v_1) < \alpha(v_2)$ and (v_0, v_2) is an edge but (v_1, v_2) is not. Because $\deg(v_0) \leq \deg(v_1)$ and $(v_0, v_2) \varepsilon E$ but $(v_1, v_2) \not\varepsilon E$, there must be another vertex v^* that is adjacent to v_1 but not v_2. Therefore v^* satisfies the remainder of (4.2) and the theorem is proved.

Therefore, rigidity in $G_R(C;\lambda)$ combined with a sequence of increasing (nondecreasing) specialization necessarily implies that species be incorporated via a perfect addition ordering. Hence natural communities with rigid circuit niche overlap graphs were built conservatively by the addition of species to individual guilds.

It should be especially obvious that this result is not affected by extinctions if the order of extinctions follows the order of arrivals. Thus the rule might apply in an ordinary successional sequence. To see that the result holds in the more general case (any elimination order) suppose it is false, i.e., that eliminating a species (vertex) results in a relatively minimal c-cutset that is not complete (species bridges more than one guild). Clearly this is not possible since in a perfect addition ordering the relatively minimal c-cutset must be complete, and eliminating a vertex in a complete graph necessarily results in another complete graph.

Also note that because species are added to single guilds does not mean that in the final niche overlap graph $G_R(C;\lambda)$ they must each belong to only one guild. Species that are added to single guilds may end up bridging multiple guilds (witness, for example, species a in Figure 4.6). It is true, however, that the necessary existence of an extreme vertex in a rigid circuit graph (Dirac 1965) guarantees the existence of at least one species in the final community that belongs exclusively to a single guild.

Apart from interpreting this rule in terms of succession or a colonization event, it may be equally reasonable to view it in an evolutionary sense. It seems reasonable to suppose that new species as they come into existence will eventually tend to compete within the single guild most closely related to its recent ancestor. If further, generalists tend to give rise to specialists the rule will be obeyed (Fitzpatrick, 1978). This does not mean, however, that communities need be strictly coevolved since there may be ecological equivalents that are capable of playing interchangeable roles in different areas. Thus each species in a particular guild could be thought of as representing a group of interchangeable species where these groups in turn bud off in evolutionary time giving rise to new groups of equivalents.

4.5 ASSEMBLY AND HOLES IN $K_C(R;\lambda^{-1})$

The rule derived above is very similar to the conservative procedure discussed in section 4.3 for building community complexes having trivial homologies. I will now show how when translated into the more detailed picture of $K_C(R;\lambda^{-1})$, the rule of adding species to single guilds guarantees the absence of holes in the resulting complex.

Suppose guilds are defined by a shared set of resources in $G_C(R;\lambda^{-1})$ or $K_C(R;\lambda^{-1})$. Addition of a new species to a single guild then involves the intersection of at least one resource v_o shared in common by members of that guild. To require that addition be made to single guilds means that any other resource v_j in the s-cutset must satisfy:

$$d(v_o, v_j) < 2$$

where $d(v_o, v_j)$ denotes the distance (minimum number of edges) separating v_o and v_j. This must be true since $d(v_o, v_j) \geq 2$ means that the resource v_j belongs to a clique other than the one defined by v_o. Therefore, all points v_j in the s-cutset must satisfy $d(v_o, v_j) = 1, j \neq 0$. This precludes the existence of one dimensional holes since clearly there must be at least one v_j such that $d(v_o, v_j) \geq 2$ to have a non bounding cycle in the s-cutset and thus a hole in the complex.

Notice that requiring $d(v_o, v_j) = 1$ is the identical constraint implemented in section 4.3 to insure homogeneity in the s-cutset. Here however, instead of postulating it as sufficient we derive it as a necessary condition from the assembly rule deduced from $G_R(C;\lambda)$.

4.6 ASSEMBLY AND INTERVALITY

In section 4.3 we saw how the above rule for conservative addition ($d(v_o, v_j) = 1$) in $K_C(R;\lambda^{-1})$ may account for the observed frequency of interval graphs in Table 4.1. The random complexes in 4.3 were constructed using empirical information on the clique sizes (simplex dimensions) for each of the species involved. In order to test more robustly whether our assembly rule is capable of explaining Cohen's finding we will apply it to the level of $G_R(C;\lambda)$ using only the total number of edges q observed as input. This is far less restrictive than using the detailed information of the clique vector (Table 3.1) as input and amounts to asking what the probability of being interval is for a random rigid circuit graph having p vertices and q edges.

Following our assembly rule, these graphs will be constructed by a random perfect addition ordering wherein each species is added to a clique of random size and all clique of a given size are equally likely to be chosen. Here species are thrust onto the parent complex with a random number of competitors within an arbitrarily chosen guild. This construction is an attempt to apply our conservative assembly rule in a manner that is as free of other structure-lending assumptions as possible. The algorithm for constructing these random rigid circuit graphs was written in the c programming language.

Numerical results are presented in section C of Table 4.2. For each of the 73 cases 1,000 random rigid circuit graphs were constructed and tested, and the frequency of those that were interval recorded here as a percentage. In all, by summing the expected probabilities estimated for each community, the number of those predicted to be interval is found to be 62.35. This agrees remarkably well with the observed number of 63 and is not significantly different according to the computed z value, $z = 0.326$. Therefore, despite its crude implementation here our very simple assembly rule appears to be strong enough to successfully predict the observed frequency of niche overlap graphs that are interval. This is significant in view of the fact that Cohen (1978) using various non-conservative hypotheses was unable to account for the apparent excess of interval graphs (nor was Yodzis, 1981).

Insofar as our procedure generates random rigid circuit graphs it appears that the rigid circuit property alone in $G_R(C;\lambda)$ is sufficient to explain the observed high frequency of interval overlap graphs. Hence the second condition of (4.1), the asteroidal property, appears to be superfluous. Therefore, rigidity in $G_R(C;\lambda)$ and the associated absence of holes in $K_C(R;\lambda^{-1})$ emerge as primary properties and intervality a secondary by-product. Cohen's discovery however, was important not only as a groundbreaking contribution to understanding the topology of niche space, but also as a signpost leading to the detection of these other properties and their common underlying assembly rule.

SUMMARY

Cohen's discovery of the ubiquity of interval food webs is shown to arise as a consequence of other fundamental topological constraints acting on communities. In particular both the absence of holes in $K_C(R;\lambda^{-1})$ and the rigid circuit property in $G_R(C;\lambda)$ are each sufficient to explain it. More fundamentally, one can use the ubiquitous rigid circuit property in $G_R(C;\lambda)$ to deduce a simple mathematical assembly rule for communities, that in turn can explain all of these topological properties. It is a simple, but necessary and sufficient rule for assembling communities that simply requires all incoming species to be constrained to compete within single guilds, as opposed to bridging multiple distantly related guilds. This simple conservative packing rule, which explains both rigidity and the lack of holes, is broadly consistent with an evolutionary interpretation for adding new species into communities.

5

HIERARCHICAL ORGANIZATION IN THE NICHE

This chapter combines the topological properties of the previous sections to show that species niches join together in a branching or tree-like pattern (see Figure 5.1). This underlying tree structure is known as *structural hierarchy*. Wangersky (1948) and Hutchinson (1957) have alluded to the possibility of this type of structure in suggesting that communities be thought of as large and loosely organized trees. Our purpose here will be to demonstrate that beyond metaphor, this form of organization is directly implied by the topological constraints already verified.

In addition, I will examine whether these constraints are consistent with the general notion of a *functional hierarchy*—functional organization involving nested groups of species defined by progressively stronger interactions. I will show how conditions consistent with an underlying tree structure in the niche also resonate with a hierarchical functional portrait for the community.

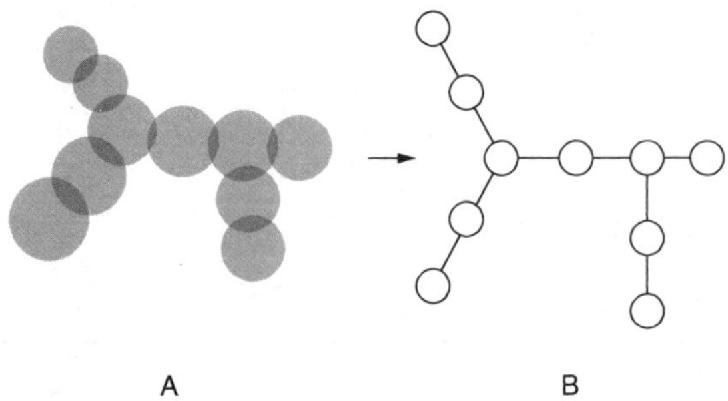

Figure 5.1 Spherical species simplexes intersect to form an underlying tree structure.

5.1 STRUCTURAL HIERARCHY

Simon (1973) defines *hierarchy* as a partial ordering among elements of a set; and it is obvious that a tree structure satisfies this definition. Thus, by identifying species as nodes (vertices) of a tree graph, a natural partial ordering is induced as follows: 1) choose any single node v_0 in the tree; 2) a partial ordering on the other vertices v_j is induced by the distance (number of edges) from v_j to v_0. In a more restricted sense Wilson (1969) describes any branched or "anastomotic" structure as hierarchical. Such hierarchy is termed *structural* to denote the fact that it is an actual topological property of system linkages (see Simon 1973; Wilson and Whyte 1969; Webster 1979; Schulman 1979 for reviews of the concept).

5.1.1 Trees: A Characterization of Rigid Circuit Graphs

We will demonstrate how the existence of a structural hierarchy in communities depends on the presence of the rigid circuit property in $G_C(R;\lambda^{-1})$ and $G_R(C;\lambda)$. The theorem explained below is a characterization of rigid circuit graphs that allows the overlap structure of a community to be collapsed into a simple 2-dimensional tree representation. This representation captures the essential pattern of species interrelationships in a community. In bare out-line the theorem shows that a tree structure is implied by an ordering on cliques similar to a perfect elimination ordering on the vertices of a rigid circuit graph. Readers not interested in the details of this theorem may skip to section 5. 2 and study Figure 5. 4 to see an example of how the result is implemented.

Theorem 5.1 (Buneman 1975)

(i) The intersection graph Ω of a set of subtrees (T_1, T_2, . . ., T_n) of a tree T is a rigid circuit graph.

(ii) A rigid circuit graph G is the intersection graph of a set of subtrees whose points correspond to maximal cliques in G such that if s_i is a vertex of T corresponding to the clique S_i of G, then the subtree T_p corresponding to vertex p in G contains s_i if and only if S_i contains p.

In short, this theorem sets up a one-to-one correspondence between the maximal cliques of a rigid circuit graph S_i and the points of a tree s_i.

Recall from section 2.2.1 that an intersection graph Ω is constructed by identifying sets as points and "non-empty intersection" as edges in Ω. Part (i) states that if the subtrees T_i are taken to be sets, that the intersection of these subtrees always forms a rigid circuit graph. This is illustrated in Figure 5.2A where if one chooses,

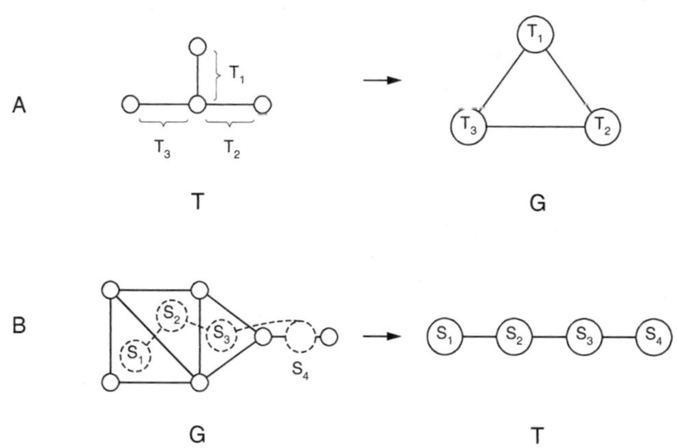

Figure 5.2 An illustration of Theorem 5.1. A) The intersection graph of subtrees T_i of a tree is rigid. B) The cliques (maximal) of a rigid circuit graph correspond to points of a tree.

for convenience, the collection of 1-simplexes as the subtrees of the 3-pointed star, the corresponding intersection graph is the triangle which of course is rigid. The result is easily proved by supposing that $\Omega(T_i)$ is not a rigid circuit graph. If this is true then by definition T must contain a non-rigid circuit (C_n; $n \geq 4$) which in turn violates the definition of a tree.

Part (ii) is slightly less straightforward to prove. As illustrated in Figure 5.2B, this part of the theorem states that for every rigid circuit graph G there is a tree T whose points correspond to the maximal cliques of G. The proof of this result parallels the argument for a perfect elimination ordering (Section 4.4) where in place of extreme vertices and c-cutsets we have *extreme cliques* and *relatively minimal cutsets*. A *relatively minimal cutset, rmc*, is the minimum number of points whose removal will isolate two given points v_0 and v_1 (Figure 5.3). Dirac (1965) has shown that every rigid circuit graph which is not complete has an *extreme clique* i.e., a maximal clique which can be disconnected from the rest of G by a complete relatively minimal cutset, *rmc*.

With these basic concepts we can sketch a proof of Part (ii) by induction on the number of cliques in G. If G is complete the result is trivial. If not, then there must be an extreme clique (Dirac) S_1 whose relatively minimal cutset is properly contained in S_1 and some other maximal clique, say S_2. Associating the points s_1 and s_2 in T with the cliques S_1 and S_2 in G, two points in T become connected *only if* they share an *rmc*. These rmc's then, give rise to equivalence relations on the maximal cliques of G. By suppressing the points in S_1 which do not belong to the relatively minimal cutset and repeating

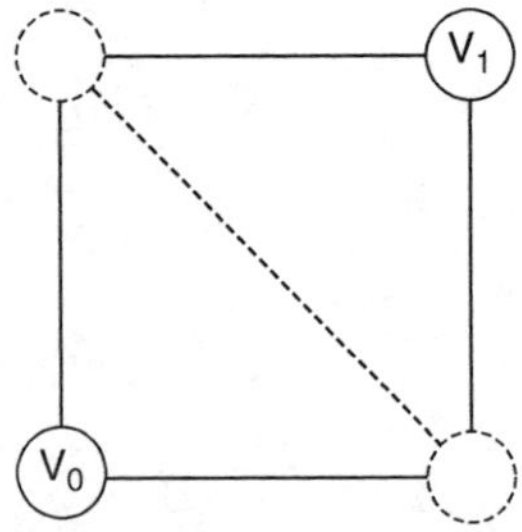

Figure 5.3 The dashed vertex pair indicates a relatively minimal cutset, in that removal of this vertex pair separates v_1 and v_0.

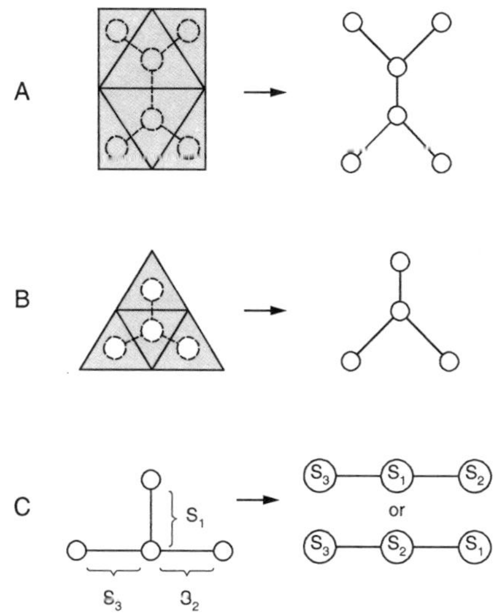

Figure 5.4 Examples of the correspondence between cliques (maximal) of a rigid circuit graph and points of a tree.

this procedure on the remaining graph, successive iterations with the concatenation of points in T_i then leads to a tree (see Buneman, 1975 for an alternative proof).

Figure 5.4 illustrates some further examples of this theorem. As can be seen in part C of this figure, the construction above does not necessarily produce a unique tree since at any stage in the induction there may be a clique which is the only member of two or more equivalence classes. The tree generated, however, is minimal in that it represents a one-to-one mapping between G and T. That is, if G is the intersection graph of subtrees of T, then there must be a map from a subset of nodes of T onto the cliques of G, and this map is one to one.

5.2 HIERARCHY IN REAL COMMUNITIES

We can now apply this result directly to $G_C(R;\lambda^{-1})$ to show how species are joined through their relatively minimal cutsets in a branching or tree-like pattern.

Recall that virtually all of the communities constructed from food web data (Table 4.1) are observed to lack holes and have niche overlap graphs $G_R(C;\lambda)$ that are rigid. A third companion property of these regularities is that the resource graph $G_C(R;\lambda^{-1})$ will have the rigid circuit property. Like triangulation in the niche overlap graph $G_R(C;\lambda)$, this property is not generic, but in randomly generated graphs it is strongly associated with a lack of holes in $K_C(R;\lambda^{-1})$. This is borne out by the fact that every one of the resource graphs $G_C(R;\lambda^{-1})$ in Table 4.1 was observed to satisfy the rigid circuit property. Moreover, the conservative packing procedure discussed in section 4.3 is able to predict this finding (72.25 rigid $G_C(R;\lambda^{-1})$ predicted; 73 observed, $z = 0.41$).

Because the resource graphs $G_C(R;\lambda^{-1})$ in Table 4.1 are rigid, Theorem 5.1 may be applied to generate the corresponding tree, denoted T(R). In order for each point of T(R) to represent a species, however, each maximal clique in $G_C(R;\lambda^{-1})$ must identify with a species' niche. As illustrated in Figure 5.5, this condition will not be satisfied if $K_C(R;\lambda^{-1})$ has holes; as some points in T(R) will fail to correspond to species. Therefore, to apply Theorem 5.1 in a meaningful way so that the points of T(R) correspond to species, $G_C(R;\lambda^{-1})$ must be rigid and $K_C(R;\lambda^{-1})$ must be homologically dense. Notice that the absence of holes *per se* does not guarantee the existence of an underlying tree structure in the niche (see Figure 5.6).

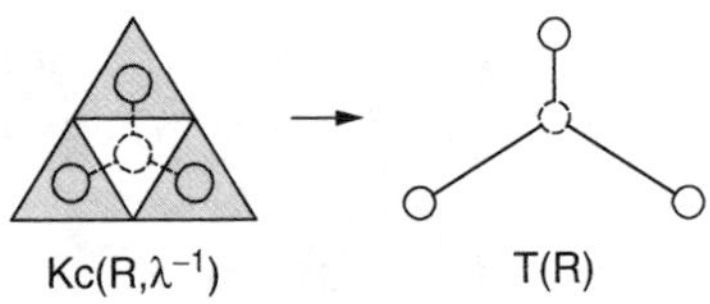

Figure 5.5 If $K_C(R;\lambda^{-1})$ has a hole, then a maximal clique in its 1-skeleton $G_C(R;\lambda^{-1})$ may not correspond to a species. The resulting tree T(R) will have a meaningless point.

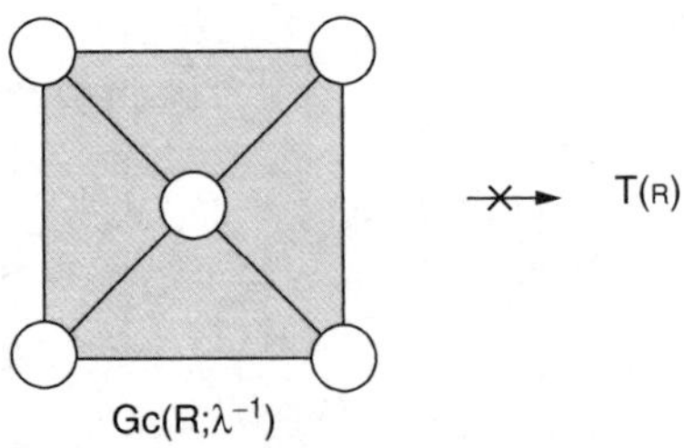

Figure 5.6 The absence of holes in $K_C(R;\lambda^{-1})$ does not guarantee the existence of a tree structure T(R) in the niche.

Another contingency in constructing T(R) is the possibility that species niches may be properly contained in the same maximal clique. This case is illustrated in Figure 5.7B. Here the cormorant's trophic niche (uses resources 12 (eel), 13 (fluke)) is contained within the merganser's niche (12, 13, 14 (blowfish)), and the kingfisher's trophic requirements (uses two kinds of minnow 15, 16) are nested in those of the green heron's (15, 16, 12). A rule of thumb for constructing T(R) in such cases is to represent the nested cliques as a star in T(R) whose hub corresponds to the generalist species (maximal

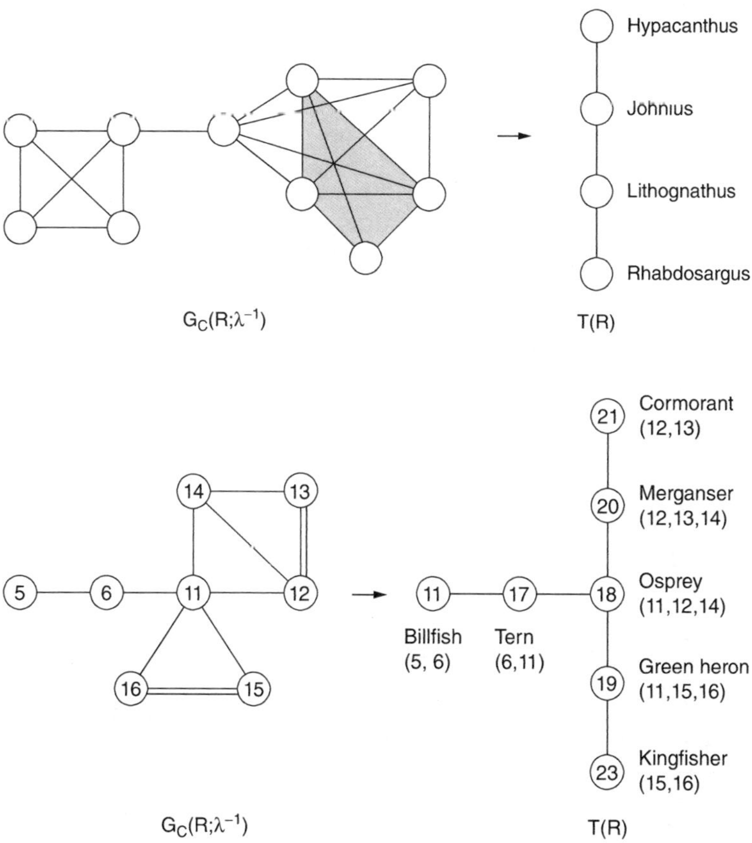

Figure 5.7 Resource graphs $G_C(R;\lambda^{-1})$ and their corresponding trees T(R) for two real communities. A) Knysna Estuary (web 2A; Day, 1967), B) Long Island Estuary (web 3A; Woodwell, 1967). See Appendix B for full legend.

clique). Clearly this modification preserves the proven one-to-one correspondence between rigid circuit graphs and trees.

We conclude therefore, that the rigid circuit property in $G_C(R;\lambda^{-1})$ and the absence of holes in $K_C(R;\lambda^{-1})$ implies that a branching or tree-like pattern characterizes the essential structure of niche overlaps in real communities. This marches with the intuitions of Wangersky (1948) and Hutchinson (1957) with regard to the structure of ecological systems.

Not surprisingly the same insight extends to $G_R(C;\lambda)$. Here, however, the maximal cliques correspond to guilds rather than species. Because real niche overlap graphs are rigid, by Theorem 5.1, guilds are also related to each other in a tree-like fashion. The corresponding tree at this level, denoted T(C), has nodes that represent guilds rather than species (see Figure 5.8). Thus the tree-like topology arises as a self-similar or scale-free property of real trophic networks.

Note that this result does not follow trivially from the analogous result on $G_C(R;\lambda^{-1})$. That is, T (R) does not necessarily imply T(C). For example, the graph on the left of Figure 5.4A, illustrates a resource graph $G_C(R;\lambda^{-1})$ whose niche overlap graph $G_R(C;\lambda)$ is not rigid (T(C) does not exist).

Because all of the real communities considered here have a tree structure on both the level of T (R) and T (C), one can say that they represent *complete* structural hierarchies. It is significant that this result would be expected if communities were constructed according to

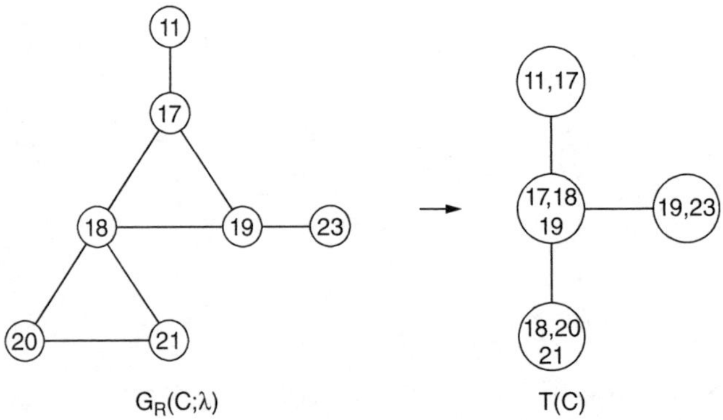

Figure 5.8 Guilds (maximal cliques) are strung together in a tree-like fashion. An example of $G_R(C;\lambda)$ and T(C) from the Knysna Estuary.

the assembly rule derived in the previous chapter. It is instructive to note that the proliferation of communities through a perfect addition ordering in $G_R(C;\lambda)$ or by conservative addition in $K_C(R;\lambda^{-1})$ can be taken to define the growth of limbs in T(C) and T(R) respectively. If one interprets this assembly process in an evolutionary sense, then T(C) and T(R) could indicate the evolutionary relationships among guilds and species in a community. This possibility, however, is not borne out by the examples in Figure 5. 7.

It is of interest to point out that if T(R) is a necessary property of natural communities, either because of constraints on their assembly or for other reasons, the presence of a hole in $K_C(R;\lambda^{-1})$ could indicate a missing species in the ensemble. Thus, for example, the hole in the communal complex in Figure 5.5 could correspond to a missing species.

5.3 FUNCTIONAL HIERARCHY

Superimposed on the binary portrait of T(R) and T(C) is an alternative characterization of hierarchy in the niche, where an ecological ensemble is seen as having a nested *functional* structure dictated by an increasing degree of functional similarity, much like a chinese box. This characterization of structure which we will call *functional hierarchy* is illustrated, for example, in a standard niche overlap dendrogram (Figure 5.9), where assemblages are broken down into smaller and more tightly related functional groups of species. This corresponds with Simon's (1962, 1973) "levels of organization hierarchy," where the systems at each level are seen to have greater interaction within subsystems than between subsystems and each functional group is defined by strong internal interaction and weaker external interaction. It is possible to generate such a picture from T(R) by eliminating edges in the underlying tree graph in order of increasing strength.[1] Disassembling the tree in this manner partitions the ensemble into smaller equivalence classes or subtrees T_i. These nested subtrees identify with the nodes in the dendrogram, D, and the terminal branches or *leaves* of D then correspond to the component species or vertices of T(R).

[1]In practice often the reverse procedure is used: concatenating stronger links first then recruiting weaker ones.

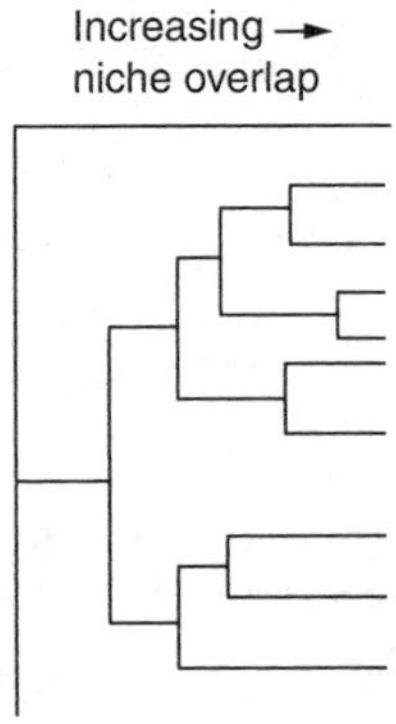

Figure 5.9 A hypothetical niche overlap dendrogram illustrating the meaning of functional hierarchy where a community is decomposed into smaller and more tightly related nested groups of species according to increasing niche overlap.

5.3.1 Compatibility Between Structural and Functional Portraits

Regardless of the topology in $G_R(C;\lambda)$ or $G_C(R;\lambda^{-1})$, a dendrogram can be always be constructed. However, some topologies allow a more consistent D than others. Consider for example the niche overlap graph and accompanying dendrogram in Figure 5.10. The edges in $G_R(C;\lambda)$ (hereafter referred to as G) are numbered from strongest (1) to weakest (6) and the dendrogram was constructed in the obvious way. Notice that although a and b are connected in G, in D they appear distantly related. In fact in the dendrogram, a is farther from b than it is from either e or d, and it is not even connected to e or d in G. Therefore the dendrogram is misleading in that the pairwise relationships shown are incompatible with the relationships contained in $G_R(C;\lambda)$. Disconnected (dissimilar) species in G are represented as *closely* related in D, while connected species are distant. As such, Simon's levels of organization concept is lost here because stronger interactions exist between branches of D than within branches.

To sharpen the discussion it is useful to introduce the notion of *compatible ordering* among graphs.

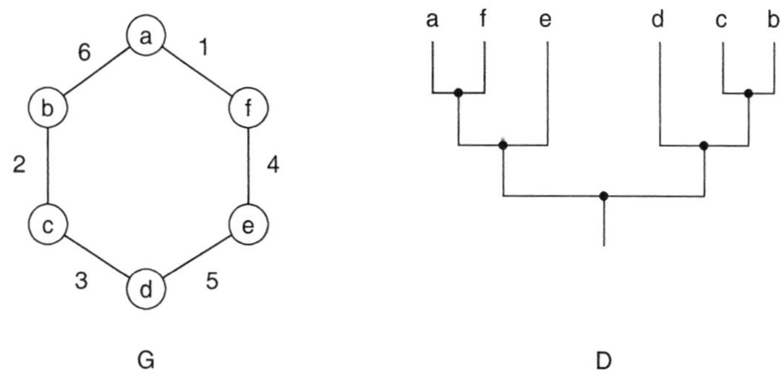

Figure 5.10 A niche overlap graph G and dendrogram D that are not compatible. Species that are closely related in G are distant in D. Here d(a, b) > d(a, e) in D, but d(a, b) < d(a, e) in G.

Definition: Two labeled graphs G and G' are compatibly ordered under the respective metrics d and d' when if:

(i) $d(v_i, v_k) > d(v_i, v_j)$ in G then $d'(v_i, v_k) \geq d'(v_i, v_k)$ in G' and if

(ii) $d(v_i, v_k) > d(v_i, v_j)$ in G' then $d'(v_i, v_k) \geq d'(v_i, v_k)$ in G for all labeled vertices v in G G'.

Let $d(v_i, v_j) = d'(v_i, v_j)$ = number of edges separating vertices i and j. Then the graphs G and D in Figure 5.10 are not compatibly ordered because d(a, b) > d(a, e) in D but d(a, b) < d(a, e) in G. The implication here is that compatibility with the niche overlap graph G may be taken to indicate how reasonably a dendrogram portrays the structure of a community.

Notice in Figure 5.10 that if edge (a, b) is eliminated from G the remaining graph is compatible. That is, compatibility is restored by breaking the non-rigid circuit. This example suggests that there may be certain properties of G, namely rigidity, that help promote compatibility in the corresponding dendrogram. Witness as a further example the similar case shown in Figure 5.11. These graphs are not compatible because d(e, c) > d(e, b) in G but d(e, c) < d(e, b) in D. However, if the non-rigid circuit a, c, d, e is broken by deleting the edge (b, e), or is rigidified with the addition of the chord (e, c) then the graph G and the dendrogram will become compatible.

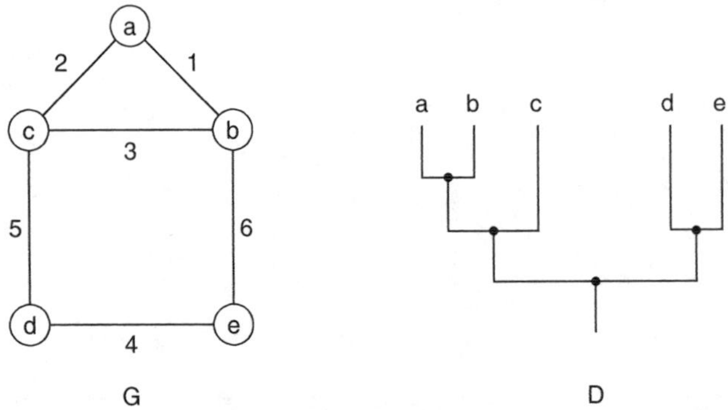

Figure 5.11 A niche overlap graph G and dendrogram D that are not compatible. They may be made compatible if G is made rigid, either by adding the chord (e, c) or by deleting the edge (a, e).

The following results summarize the relationship between rigidity in G and compatibility with D.

(i) For every dendrogram D there exists a compatible rigid niche overlap graph G.

Proof: The complete graph is rigid and is compatible with every dendrogram.

(ii) Part (i) is not true for non-rigid circuit graphs.

Proof: See Figure 5.12 for counterexample.

(iii) Every rigid circuit graph G, $p \leq 6$, not containing the star $K_{1,n}$, $n > 3$, as an induced subgraph (a subset of vertices with all the edges as they exist in G) has a compatible dendrogram.

Proof: By exhaustion of possibilities for p, q-graphs in Harary (1972, Appendix A).

(iv) Part (iii) is not true for non-rigid circuit graphs.

Proof: Figure 5.10 is a counter-example.

Notice that (iii) does not necessarily mean that any specific dendrogram will be compatible with G, only that one exists. It is possible that a particular ordering on the edges of G will not yield a compatible dendrogram even though G itself is rigid. If G is complete, however, then every possible dendrogram will be compatible.

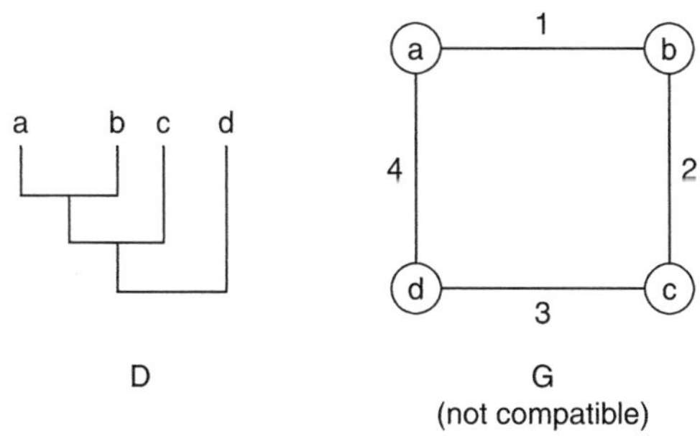

Figure 5.12 A dendrogram D for which there is no compatible niche overlap graph G.

The results above demonstrate that the topological structure of $G_R(C;\lambda)$ does in fact influence whether or not an assemblage can be meaningfully represented as a functional hierarchy. In particular they suggest that compatibility is more likely to be obtained if G is rigid than if it is not. Therefore, the existence of structural hierarchies in real communities marches with the possibility that these systems are organized as compatible functional hierarchies. This is consistent with Simon's (1962, 1973) levels of organization concept where systems are organized as nested sets with each level defined by stronger internal interactions than external ones.

SUMMARY

Real niche spaces are shown to be organized into structural hierarchies where species niches are joined in a branching or tree-like pattern. This structure is consistent with the assembly rule derived in the last chapter since a perfect addition ordering or conservative assembly procedure will give rise to an underlying branching pattern in the niche. Such topology in niche overlaps supports the view that communities are organized into functional hierarchies where nested subgroups are defined by progressively stronger interactions. In the following chapter we will see how this form of organization helps to stabilize communities, making them more persistent and more resilient to outside perturbations.

6

DYNAMIC LINKAGES AND TOPOLOGICAL PATTERNS

Up to this point we have focused on the topological structure of niche space and have demonstrated how the static patterns we uncovered might be underpinned by simple constraints on how the systems were assembled. In this chapter we will begin to look at an alternative facet of the problem and examine the possible dynamic causes and effects of the structural regularities. Our aim here will be to demonstrate in a preliminary way how the topology found in real communities may be selected for and may resonate with increased stability in natural ecological systems. Note that mathematical stability implies to first approximation that a system will be more persistent (more likely to be observed), whether or not it is correct to describe nature as stable.

This contributes to a large and growing literature seeded by May (1972) on the "special tricks" employed by natural systems that reconcile dynamic stability with the improbability of this circumstance in randomly constructed model communities. In this regard, particularly important mechanisms have been suggested by DeAngelis (1975) and Nunney (1980) (see May 1979 for a recent review). Here, we will look directly at the structural regularities uncovered in the previous chapters to see if these place real ecological ensembles in the relatively narrow regions of parameter space possessing enhanced stability-persistence properties.

To begin, we will focus on rigidity in the niche overlap graph and its effect on the competitive dynamics of communities. Restricting our view to this level is in part justified by the fact that the rigid circuit property in $G_R(C;\lambda)$ is tightly associated with structural hierarchy and the absence of holes in $K_C(R;\lambda^{-1})$. Moreover, these topological patterns do not translate to patterns in the deeper predator-prey picture of food webs (see Figure 6.1), but appear to be suspended on the skin of the competitive relationships in the community. While this makes things simpler it is unfortunate that the body of techniques for the dynamic analysis of directed graphs cannot be brought to bear here (e.g., Chen, 1972; Harary, 1959; Parter, 1960; Roberts, 1974; Levins, 1974).

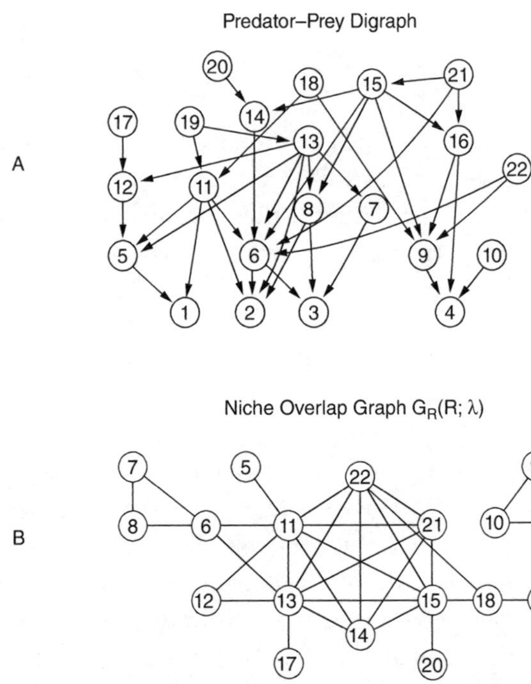

Figure 6.1 A predator-prey digraph and niche overlap graph $G_R(C;\lambda)$ for the Whytam Wood community Web 27 (Varley, 1970; see Appendix B for legend).

6.1 TRIANGULATION AND STABILITY: A SIMPLE EXAMPLE

We begin by examining how rigidity in $G_R(C;\lambda)$ can affect the stability of a system. As a preliminary step we will examine the competitive dynamics of generalized *nonlinear* systems near equilibrium using the standard linear constant model for S interacting species described by the differential equation

$$\dot{x} = \underline{A}x \tag{6.1}$$

where $x = x(t)$ is an S-vector $x = \{x_1, x_2, \ldots, x_S\}$ the components of which represent the magnitudes of the perturbations of species populations from equilibrium. The S × S constant community matrix *A* has elements α_{ij} which characterize the influence of the jth species on the ith species in the community. All main diagonal elements are assumed to have a value of −1 to reflect the usual intraspecific resource limitation (May, 1973). To study the effects of structure on the stability of the equilibrium $x_e = 0$ in model 6.1 we will assume the elements α_{ij} of $\underline{A}$ have the form

$$\alpha_{ij} = \begin{cases} -1, & i = j \\ e_{ij}\, a_{ij}, & i \neq j \end{cases}$$

where $a_{ij} = \alpha_{ij}$ before structure is added to the matrix and the numbers e_{ij} are the elements of a symmetric adjacency matrix that defines the edges of $G_R(C;\lambda)$. That is, $e_{ij} = 1$ or 0 depending on whether or not an edge exists between vertices (species) i and j in $G_R(C;\lambda)$.

Consider the example illustrated in Figure 6.2 of two alternative niche overlap graphs. These systems are identical in that each has

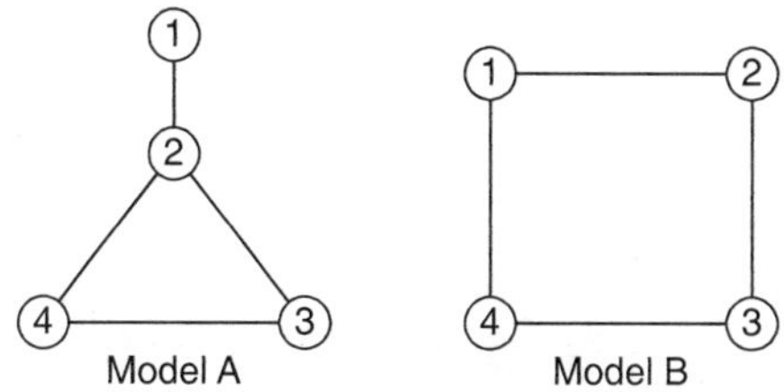

Figure 6.2 Two hypothetical niche overlap graphs having the same number of species and edges.

four species (vertices) and the same connectance (c = 0.67), however they differ in that model A is rigid but model B is not. The corresponding community matrices *A* for the systems are:

Model A

$$\underline{A}_A = \begin{bmatrix} -1 & \alpha_{12} & 0 & 0 \\ \alpha_{21} & -1 & \alpha_{23} & \alpha_{24} \\ 0 & \alpha_{32} & -1 & \alpha_{34} \\ 0 & \alpha_{42} & \alpha_{43} & -1 \end{bmatrix}$$

Model B

$$\underline{A}_B = \begin{bmatrix} -1 & \alpha_{12} & \alpha_{13} & 0 \\ \alpha_{21} & -1 & 0 & \alpha_{24} \\ \alpha_{31} & 0 & -1 & \alpha_{34} \\ 0 & \alpha_{42} & \alpha_{43} & -1 \end{bmatrix}$$

To gain an initial idea of which of these structures has a larger stable parameter space we will set $\alpha_{ij} = a$ and solve routinely for the corresponding eigenvalues λ. The coefficients for the characteristic polynomial:

$$\lambda^4 + c_1\lambda^3 + c_2\lambda^2 + c_4 = 0$$

for model A are: $c_1 = 4$, $c_2 = 6 - 4a^2$, $c_3 = 4 - 8a^2 - 2a^3$, $c_4 = 1 - 2a^3 - 4a^2 + a^4$; and for model B: $c_1 = 4$, $c_2 = 6 - 4a^2$, $c_3 = 4 - 8a^2$, $c_4 = 1 - 4a^2$. Instead of evaluating λ's directly it is convenient here to use the *Routh-Hurwitz* stability criterion (Stewart, 1960; May, 1974). The explicit necessary and sufficient conditions for stability in the four-species case are:

$$c_1 > 0,\ c_3 > 0,\ c_4 > 0$$
$$c_1c_2c_3 > c_3^2 + c_1^2c_4$$

For $-1 \leq \alpha \leq 0$ the sensitive condition is $c_3 > 0$ for model A and $c_a > 0$ for model B. It is easy to see that the size of the stable parameter space for the triangulated model A ($-0.80 < a < 0$) is greater than for the non-rigid community B ($-0.5 < a < 0$). That is, c_4 for model A is greater than c_4 for model B and is positive over a slightly larger range of a.

To show that this is a robust result we will relax the assumption that α's are equal and assume instead that each α_{ij} is chosen randomly from a uniform distribution between −1 and 0. In this case, the fraction of 1000 Monte Carlo replicates computed to be stable for each of these models is:

$$\text{fraction A stable} = .892$$
$$\text{fraction B stable} = .520$$

Therefore, community A, which is identical to community B in all respects except with regard to its triangulated overlap structure, is more than 50% more likely to be stable than B. This limited but promising result suggests that rigidity in $G_R(C;\lambda)$ may work to enhance the stability-persistence of a system. Again, one is more likely to observe a system that is persisting.

The outcome of this example is somewhat surprising when viewed from the coarse perspective of the *Gerschgorin* criterion (Stewart 1960). This theorem characterizes the eigenvalues λ_i of an $S \times S$ matrix *A* as lying within S discs D_i in the complex plane whose centers each correspond to the value of the ith main diagonal element, and whose radius is equal to the absolute value of the ith row sum, i.e.,

$$D_i = \left\{ \lambda : \left|\lambda - \alpha_{ii}\right| < \sum_{\substack{j=1 \\ j \neq i}}^{S} \left|\alpha_{ij}\right| \right\} \qquad i = 1, 2, 3, \ldots, n$$

In the above example, all discs are centered at (−1, 0) in the complex plane. Species that interact with a greater number of other species will have discs of larger diameter. Hence, a vertex in $G_R(C;\lambda)$ of high degree will in general correspond to an eigenvalue that falls within a Gerschgorin disc of larger diameter. These larger discs in turn will have an increased probability of crossing over into the positive half-plane. Therefore, a species that interacts with a greater number of other species will have an increased chance of associating with an eigenvalue *with a positive real* part.[1] Because species 2 of model A has more neighbors than any species in model B, system A may be expected to be less stable than B. However, because A is in fact more stable than B something more subtle must be at work here than can be predicted from the point degree of these systems. In the next section we will see that this effect may be related to the lower stability of non-rigid circuits.

6.2 TRIANGULATION AND STABILITY: DYNAMIC SELECTION

The example above considered the stability of the equilibrium without regard to its *feasibility*. That is, without the added constraint that the equilibrium be positive (Roberts, 1974; Gilpin, 1975; Strobeck, 1973). Here we extend our results with a numerical study of feasible equilibria for the general case of rigid versus non-rigid circuit graphs.

Tregonning and Roberts (1979) have recently advanced the intriguing hypothesis that stable communities may come into being by a process of dynamic decay whereby a large multi-species ensemble collapses to a smaller stable one through a sequence of extinctions. They show that the extremely low likelihood of obtaining a stable feasible equilibrium *denovo* ($< 2^{-S}$) is greatly enhanced by allowing an assemblage to arrive at a given size by winnowing down through a process of dynamic selection. Not only is this biologically plausible, but from a technical standpoint it is the most practical way to study rare feasible equilibria.

In the present context, dynamic selection has special significance in having the potential to shape certain topological characteristics of communities. For example, if non-rigid circuits in $G_R(C;\lambda)$ are found to be less stable and are selectively eliminated, the reduced feasible communities should contain an excess of triangulated overlap graphs. If the effect is strong enough dynamic stability may be invoked as

[1]Assuming density is uniform in each disc the asymptotic probability for S large that any species will have an eigenvalue with a positive real part approaches 1/2 from above. The probability that a system with S species is stable may then be estimated as the product of the fraction of the area of each disc in the negative half-plane. For a large random system this approaches 2^{-S}. Notice the correspondence here with Goh and Jennings' (1975) results on feasibility.

a partial cause of the topological patterns that are observed. This is reinforced by the fact that eliminating points in rigid circuit graphs always leads to other rigid circuit graphs (Dirac, 1965), whereas suppressing points in non-rigid graphs may cause them to become rigid (Figure 6.3). The rigid circuit state is a capturing state—"once rigid, always rigid."

Tregonning and Roberts (1979) studied dynamic decay by sequentially eliminating species that, by an intuitive criterion, were thought most likely to go extinct. Here rather than guessing at the extinction order we will simulate the decay process brute force. The dynamics governing this decay will be modeled using the standard finite difference analogue of the Lotka-Volterra competition equations.

$$N_i(t+\tau) = N_i(t) + \tau r_i N_i(t)\left[K_i + \sum_j \alpha_{ij} N_j(t)\right] \tag{6.2}$$

where $N_i(t)$ is the population size of species i at time t, r_i is the intrinsic rate of growth without density dependence, K_i is the carrying capacity, τ is the time increment, and α_{ij} is the effect of species j on species i.

The values used in the simulations are as follows: $N(0) = 1$, $r_i\tau = 0.001$, $K_i = 1000$. The value of $r_i\tau$ was chosen as 1/10 the value at

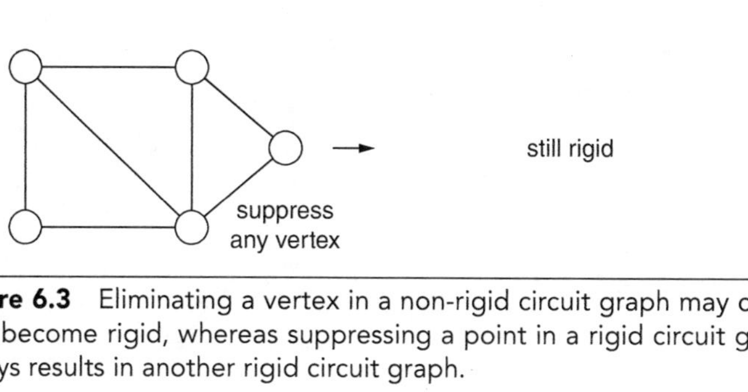

Figure 6.3 Eliminating a vertex in a non-rigid circuit graph may cause it to become rigid, whereas suppressing a point in a rigid circuit graph always results in another rigid circuit graph.

which cycling was first observed. No runs using this increment were observed to cycle. The coefficients α_{ij} were chosen randomly from the interval [−0.1, − 0.9], then normalized by the row sums of the adjacency matrix so that species that compete with many others will do so more weakly than species that have few competitors (Harrison, 1979; Hastings, 1982). That is,

$$\alpha_{ij} = a_{ij} \frac{\left(\min_{i=1,n} \sum_j e_{ij} \right)}{\sum_j e_{ij}}$$

where a_{ij} is chosen from the uniform distribution between −0.1 and −0.9.

The adjacency matrix for the initial community was generated to correspond to a random connected graph (cf. chapter 4) with 15 vertices and 53 edges (c = 0.5). These random communities of 15 species were then set into motion with $N_i(0) = 1$ and allowed to grow and decay into the final feasible communities. Extinction occurred for species i when $N_i(t) = 0$ whereupon N_i was set equal to zero for all t thereafter. Each run was terminated when:

$$\left(N_i(t+\tau) - N_i(t)\right)^2 < 0.01 \quad , \text{ for all i.}$$

This appeared to be conservative as all runs stabilized at a particular set of integer values for N_i long before termination occurred.

Results of twenty simulations, beginning with a random community of fifteen species, are shown in the first two columns of Table 6.1. Interestingly, eighteen out of twenty feasible communities were rigid. On the surface, it appears that dynamic selection may favor the generation of triangulated niche overlap graphs. In closer view, however, the high frequency of rigidity may simply be an artifact of generating relatively small final communities. To examine the significance of these results, we will test them against two alternative null hypotheses. Each examines the possibility that the resulting feasible communities were generated at random without preference for rigidity.

In the first null hypothesis we look at the largest connected component derived from each decay. In all cases this was the only nontrivial component generated, i.e., having more than two vertices. We

Table 6.1 Dynamic Decay Model

Dynamic Decay p* Species Surviving	Rigid?	Expected Pr Largest Component is rigid	Expected Pr Rigid w/ Random Elimination
7	1	1	0.70
9	1	0.37	0.52
7	1	1	0.70
7	1	0.62	0.70
6	1	1	0.86
8	1	0.48	0.63
7	1	1	0.70
6	1	1	0.86
8	0	0.22	0.63
7	1	0.57	0.70
7	1	0.64	0.70
7	1	0.64	0.63
8	0	0.45	0.63
7	1	1	0.70
8	1	0.62	0.63
6	1	1	0.86
9	1	0.64	0.52
7	1	0.64	0.70
6	1	1	0.86
8	1	0.62	0.63
Totals:	**18**	**14.51**	**13.96**
		z = 2.04	**z = 2.09**

then determine the probability that this component would be triangulated if it was constructed as a random connected graph (cf. Chapter 4). For each case the observed numbers of points and edges are used to estimate the probabilities shown in column 3 of Table 6.1 (1,000 replicates each). Summing these probabilities, it is found that if the components were generated at random 14.51 would be expected to be rigid. This differs significantly from the 18 produced by dynamic decay. The computed z-statistic here is $z = 2.03$, where once again, $z \geq 1.65$ indicates significant difference at the 5% level.

Similar results are obtained if one looks at the consequences of eliminating species randomly, versus by extinction in dynamic selection. Here we determine the probability that a given feasible assembly of p species is rigid if it was generated by eliminating 15 – p vertices at random. Results are presented in column 4 of Table 6.1. By summing probabilities, the expected number of rigid overlap graphs that would have been produced by sequential random elimination is found to be only 13.96. This again differs significantly from the 18 generated by dynamic selection, $z = 2.09$.

These results indicate that the process of dynamic decay has a tendency to selectively eliminate species involved in non-rigid circuits. Therefore communities having non-rigid overlap structures will have a lower likelihood of being stable than those having rigid niche overlaps.

This point is reinforced in Table 6.2. Here we compare the number of feasible species surviving the decay process as a function of structure in the initial community. In decay 1, the initial community of 15 species is free of structure in that it is constructed randomly as before. In decay 2, it is constructed to be rigid by using a perfect addition ordering. The starting number of vertices ($p = 15$) and edges ($q = 53$) is the same in both cases. As shown in Table 6.2, corresponding to the predictions above, triangulated communities are more likely to be stable in having fewer extinctions ($\bar{x} = 5.3$) than the randomly constructed communities ($\bar{x} = 7.8$). By comparing variances, this difference can be seen to be significant. Therefore, insofar as real niche spaces are rigid, they should possess a higher likelihood of being stable.

Although it is tempting, it would be misleading to attempt to apply these results in the other direction, to explain observed topological patterns. In particular, one should not interpret Table 6.1 too literally since the frequency of triangulated niche overlap graphs obtained is sensitive to initial conditions. For example, when the initial random graphs of 15 points were generated to have 95 edges instead of 53, three out of five final communities were observed to be non-rigid. These graphs had more independent non-rigid circuits than could be effectively broken by dynamic decay (recall from Eq. 3.5 Rank (cycle basis) $= q - p + 1$). In defense, however, I should add that the parameter values chosen were thought to be reasonable. In particular, the number of edges used in the initial graphs (53) was chosen to yield

Table 6.2

Trial I.D.	Decay 1: Random Initial Community p* Species Surviving	Decay 2: Triangulated Initial Community p* Species Surviving
a	7	8
b	9	11
c	7	9
d	7	11
e	6	9
f	8	11
g	7	10
h	6	8
i	8	10
j	7	10
	Avg. surviving = 7.2	**9.7**
	σ = 0.92	**1.15**
$\bar{p}$* Decay 1 < $\bar{p}$* Decay 2		

Size of feasible equilibrium p* under dynamic decay for random versus triangulated initial communities. Starting number of species = 15, initial edges = 53. Triangulated initial communities have more survivors. 10 trials.

a connectance (0.5) that agrees roughly with empirical values (Table 4.1). Nonetheless, because this sensitivity exists, it may be unwise to generalize the predominance of rigid circuit graphs obtained in Table 6.1 in a literal sense to try to explain the genesis of natural topological patterns.

In addition, it should be kept in mind that we are dealing here with a limited picture by focusing on competition and ignoring predator-prey dynamics. This is especially hard to justify when transcending trophic levels. The spirit in which this analysis was performed was to determine in caricature the effects of such structure in niche overlaps *per se* on the stability of communities. It would be a mistake therefore to try to extend them literally to generate the observed topological patterns in Table 3.1.

Therefore, rather than attempt to attribute an ultimate dynamic cause to these regularities, it seems more reasonable to view them in terms of the constraints already discussed on the homogeneity of species niches and conservative assembly. Certainly these constraints may have some dynamic background but it seems too harsh to view them simply as dynamic by-products. It is lunatic, for example, to think that fish avoid competing with birds in the upper canopy for dynamic reasons. On the other hand, I believe it is safe and constructive to say that the topology of real niches helps to place natural communities in the sparse regions of parameter space possessing enhanced stability-persistence properties.

6.3 HIERARCHY AND STABILITY

Here we briefly mention two additional aspects of structure that may work to enhance stability: the advantages of structural hierarchy and functional hierarchy.

6.3.1 Structural Hierarchy

May (1972) and Wigner (in Mehta, 1967) have proposed the following asymptotic criterion for stability that is a profound ensemble extension of the Gerschgorin theorem (Hastings 1982). A random $S \times S$ community matrix $\underline{A}$ is stable for S large if:

$$\alpha(Sc)^{1/2} < \langle \mathrm{tr}\ \underline{A} \rangle \tag{6.3}$$

where α is the standard deviation of non-zero elements of $\underline{A}$, S is the number of species, c is connectance, and $\langle \mathrm{tr}\ \underline{A} \rangle$ is the absolute value of the mean of the main diagonal elements of $\underline{A}$. When the quantities α, S and c are treated as independent, we arrive at the well-known inverse relationship between complexity and stability (Gardner and Ashby 1970; May 1972). That is, the larger and more complex a randomly constructed dynamic system becomes, the less likely it is to be stable. On the other hand, if any of these quantities are negatively coupled, then the inverse relationship between complexity and stability may break down. Here we will show how such negative coupling between numbers of species, S, and connectance, c, follows from our results on the existence of structural hierarchies in nature. This analysis marches with May's (1972) early insight that nature may be organized as loosely coupled subsystems.

Recall that on the level of niche overlaps, most natural communities can be portrayed as a network of guilds strung together in a branching or tree-like fashion. That is, $G_R(C;\lambda)$ can be characterized as a tree graph T(C) whose vertices correspond to guilds or maximal cliques in $G_R(C;\lambda)$.

Because the connectance of a tree is simply:

$$c = \frac{q}{\binom{p}{2}} = \frac{2}{p}$$

the connectance of a community organized as a structural hierarchy can be computed roughly as:

$$c = \frac{\xi(p-1) + \binom{D}{2}p}{\binom{S}{2}} \tag{6.4}$$

where S is the number of species, p is the number of guilds (vertices of T(C)), D is the average size of a guild and ξ is the average size of a relatively minimal cutset joining the guilds. For D and S large, this reduces to the approximation:

$$c \cong \frac{D}{S} \tag{6.5}$$

Thus, in a structural hierarchy for a given guild size D, connectance c should scale inversely with the numbers of species S. Interestingly, such inverse coupling between c and S has been observed decisively in a number of important empirical studies by McNaughton (1978) and Rejmanek and Stary (1979). Yodzis (1980), however, observes a slightly nonlinear inverse relation for Cohen's (1978) food web data.

For structural hierarchies the stability criterion 6.3 then becomes:

$$\alpha D^{1/2} < \langle \mathrm{tr} \quad A \rangle. \tag{6.6}$$

This implies that for dynamic reasons real communities should proliferate not by increasing the *size*, D, of existing guilds, but by increasing the *number*, p, of different guilds. Figures 6.4 and 6.5 bear this out for the corrected Briand communities (connected components)

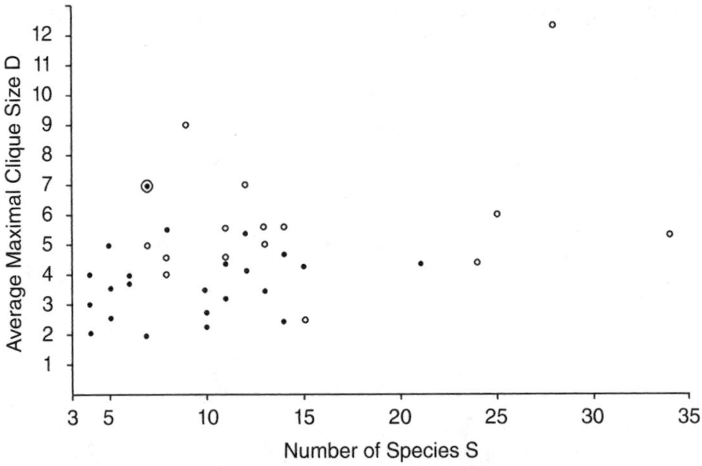

Figure 6.4 A plot of average maximal clique size D in $G_R(C;\lambda)$ against numbers of species S for the communities in Table 4.1. For $S > 4$, $D = 0.095\ S + 3.6$, $r = 0.32$. Therefore natural communities attain large sizes by increasing the size of guilds (maximal cliques).

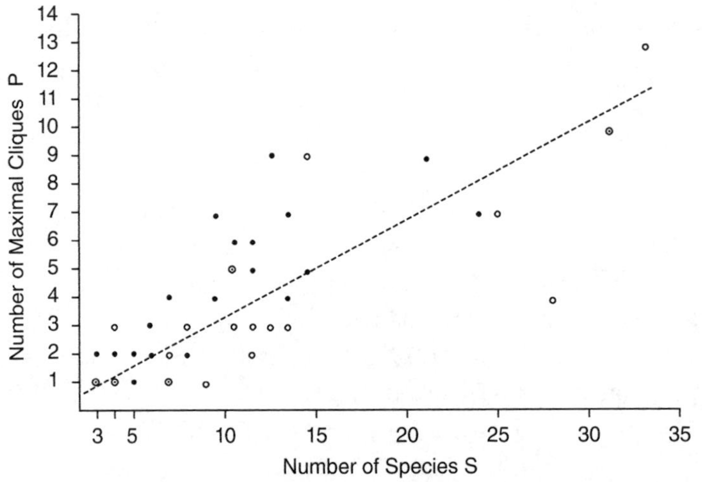

Figure 6.5 Numbers of maximal cliques p in $G_R(C;\lambda)$ plotted against numbers of species S for communities in Table 4.1. For $S \geq 3$, $D = 0.31\ S + 0.46$, $r = 0.80$. Therefore natural communities tend to attain large sizes by increasing the number of guilds. Note that the inverse of the slope should roughly equal the size of an average guild (3.2 ~ 3.6).

in Table 3.1, where $S > 2$. Here there is no clear relationship between S and D, however, there is a definite and positive relationship between S and p. Of special interest, therefore, is that these real communities tend to proliferate more by creating new guilds than by increasing the size of existing guilds (maximal cliques).

This rule of growth agrees nicely with the increasing specialization argument used earlier in deriving our assembly rule. If each additional species is more of a specialist (lower point degree in $G_R(C;\lambda)$ than any of the established species, rather than joining to a maximal clique in $G_R(C;\lambda)$ it will tend to attach to a submaximal clique. Such submaximal attachment will necessarily generate a new maximal clique or guild. According to the simple dynamic argument above, this resulting structural hierarchy is more likely to be stable than if new species join completely with the largest clique. Therefore natural communities may be able to avoid the stability-complexity barrier by being organized as structural hierarchies generated by the conservative assembly rule that was derived in Chapter 4.

Notice that equation 6.6 also implies that if communities experience structural disturbances that alter the values of the interactions among species, they should collapse more frequently by decreasing the size of existing guilds rather than the number of guilds. Therefore insofar as the tropics may act as a species pump for the temperate zone (Darlington, 1957; Haeffer, 1969) one would expect the thinning out of species as one moves away from the tropics to be characterized more by a reduction in size of guilds, D, than in number of guilds, p. This is borne out in a preliminary study of guild structure in bird communities by Terborgh (1980).

6.3.2 Functional Hierarchy

In section 5.4 it was argued that the existence of a structural hierarchy increases the chances of being able to decompose a community into a meaningful functional hierarchy. This form of organization makes it possible to separate a community into nested groups where each level is defined by stronger internal interactions than external interactions.

Siljak (1975) and Goh (1979) have demonstrated that the degree to which such hierarchical decomposition is possible may greatly enhance the stability properties of a system. They use an aggregation-decomposition analysis to show that if from the total system-view the interactions within subsystems are stronger than the interactions between subsystems, then the system can be stable even with strong interactions or with time-dependent changes in the interactions among components. This is an area which I believe has a great deal of promise and to which a diakoptic or partitioning and tearing analysis might be fruitfully applied to take advantage of the structural regularities found here (Kron, 1963; Stewart, 1965; Parter, 1960; Harary, 1976). I leave this as a future project.

SUMMARY

This chapter considers various ways in which topological structure can enhance the stability properties of real communities. It begins by showing how rigidity in niche overlaps provides a circuit structure that is more likely to lead to stable and feasible equilibria than when larger non-rigid circuits prevail. Hence rigidity in $G_R(C;\lambda)$ as a companion property of homological density in $K_C(R;\lambda^{-1})$ tends to place real communities in a special corner of parameter space possessing enhanced stability properties. On a coarser scale, it is shown how organization as a structural hierarchy helps to by-pass the stability-complexity barrier. Such structure puts simple constraints on the connectance of real communities that enable them to defend against the sharp decline in stable equilibria characteristic of random systems. This marches with May's (1972) suggestion that nature may be organized as loosely coupled subsystems. Finally, the work of Siljak (1975) and Goh (1979) is pointed out to enforce the view that insofar as real communities are organized as functional hierarchies their stability properties should be enhanced. Thus the initial results from Chapter 5 on compatibility and structure, may have some relevance to the dynamic behavior of natural systems. And if so the dynamic behavior may feed back as part of the selective mechanism behind the assembly rules of previous chapters.

7

HIERARCHIES OF ABUNDANCE

Our aim in this chapter will be to demonstrate a correspondence between two classical ideas of structure in natural communities: structure in a functional sense as in Simon's "levels of organization" concept (Chapter 5), and structure as manifested in patterns of species relative abundance.

Early attempts to find mathematical regularity in species abundances include the biogeographical works of Jaccard (1908), Arrhenius (1921) and Gleason (1922). However, it was not until Motomura (1932) and Fisher, Corbet and Williams (1943) came forth with their logseries model for the Malaysian butterfly collections that quantitative studies of species abundance really came into being. Since then, the field has expanded rapidly sending forth a number of approaches, ranging from resource apportioning theories and statistical descriptions to a variety of diversity indices, some rooted in the mythology of an information theoretic paradigm (see May, 1975; Pielou, 1975; Patrick, 1982 for reviews). An underlying motive for all of these studies was to find some characterization of species abundances that could be related meaningfully to the functioning and organization of communities.

In previous chapters it was argued that the topology of real communities is consistent with the existence of an underlying functional hierarchy. Here, we will examine the consequences of such organization for patterns of species abundance. By so doing we hope to contribute at least in part to the motivation for the scientific study and interest in species diversity: the idea that through such readily obtained community statistics as relative abundances, one can gain insight into the functioning and organization of communities.

A model has recently been proposed which attempts to understand relative abundance patterns in terms of a hierarchically structured niche (Sugihara, 1980; Appendix D). In the section below we will briefly summarize the main ideas and principal results of this work, and will introduce new findings which reinforce and elaborate the proposed correspondence between abundance patterns and functional organization in the niche.

7.1 THE NICHE HIERARCHY MODEL

Consider the overlap dendrogram in Figure 7.1. The community is portrayed here as a functional hierarchy consisting of nested sets of smaller and more tightly related functional groups of species. The lower bifurcations in the dendrogram define the weaker interactions

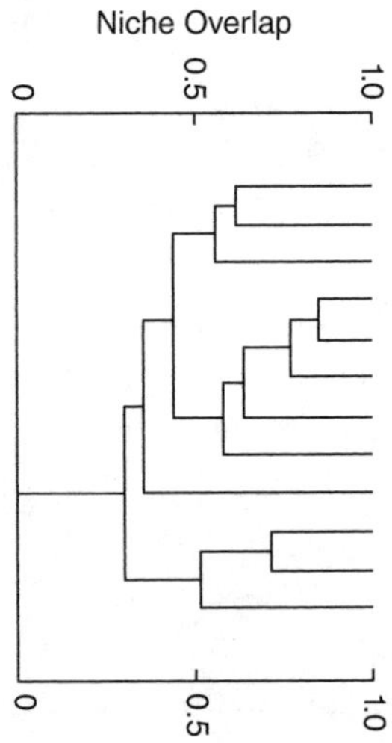

Figure 7.1 A hypothetical niche overlap dendrogram illustrating functional hierarchy in communities. In the niche hierarchy model, each bifurcation in the functional portrait corresponds to a subdivision of the communal resource pie.

while the higher bifurcations correspond to stronger and more robust linkages among species. In essence the dendrogram defines a sequence by which the niche could be peeled apart, beginning with the weakest segments of the community on up to the most strongly connected groups of species. Translated into the picture of the resource complex $K_C(R;\lambda^{-1})$ or the niche overlap graph $G_R(C;\lambda)$, the dendrogram may be thought of as ranking the strength of cleavage planes in the niche defined by *point cutsets* in $K_C(R;\lambda^{-1})$ or the resource graph $G_C(R;\lambda^{-1})$ and *line cutsets* in the consumer graph $G_R(C;\lambda)$.

In its essential outlines the niche hierarchy model views relative abundance patterns as a direct expression of this functional portrait of the niche. It investigates the obvious relative abundance distribution that might arise from a niche topology that allows communities to be naturally subdivided into smaller functional groups of species. To a first order, a physical metaphor for this process would be to liken the communal niche to a unit mass or multidimensional crystal which is sequentially split apart to yield fragments that denote relative species abundance. This simple and grossly phenomenological conception is in the spirit of the MacArthur (1957) broken stick model with the exception that breakages here are sequential and multidimensional. A given dendrogram may be thought of as a tracing of the sample path of a sequential breakage process that divides a communal resource pie among component species where with each slice or subdivision may involve different sets of niche factors.

Kolmogoroff (1941) has shown that such sequential fractures will lead asymptotically to lognormal size frequency distributions. The breakages themselves do not have to be random, however, the magnitude and frequency of breakage must be independent of particle size. Thus, for example, gravel fragments resulting from repeatedly crushing rocks often tend to be lognormal. Instead of smashing rocks one can imagine dividing up relative species abundance in a way that reflects a hierarchically organized niche. Therefore, large numbers of ecologically related species should tend to have a lognormal distribution of abundance. This agrees with an extensive amount of empirical evidence showing that lognormal distributions obtain in virtually all large collections of organisms including diatoms (Patrick, 1968; Patrick et al., 1954), soil arthropods (Hairston and Byers, 1954), lepidoptera, birds, and mammal faunas (Preston, 1962), and for areas ranging in size from diatometer slides to entire continents. As an explanation for the genesis species of gross relative abundance patterns, the niche hierarchy model differs from other classical hypotheses in that it produces a lognormal curve possessing a stable variance. The principal shortcoming of the other seemingly sensible explanations of lognormality as the product of continual random multiplicative effects on growth is that they imply an ever increasing variance in the distribution with time (Sugihara, 1980).

Given that abundances are divided sequentially, to make meaningful predictions from the model one needs to determine how they are divided at each step. That is, one needs to determine the breakage distribution F (Appendix D) applied at each step in the model. In biological terms this amounts to asking whether apportionment is one-dimensional and uniform, or multidimensional where the final outcome is an average of random apportionments on several niche dimensions. To determine this empirically, data from two-species associations of fish and barnacles were considered (Figure 7.2). Here it was found that the distribution of proportional abundances tended to cluster into a triangular pattern rather than spread out into a flat uniform distribution. This appears to support the multidimensional hypothesis. Furthermore, when this rule is extended to the three-species case it produces very good agreement with the data.

The most significant characteristic of the niche hierarchy model is that it is able to generate the specific one-parameter class of *canonical lognormal* distributions that actually occur in nature (Preston, 1962). These distributions are "canonical" because their parameters tend to be coupled in the following way:

$$S = \sigma(\pi / 2)^{1/2} \exp\left[\frac{(\sigma \ln 2)^2}{2}\right] \tag{7.1}$$

where S and σ^2 are the species count and variance of the lognormal assemblage. Therefore according to the canonical hypothesis species count and variance are positively related. This has special importance to the theory of island biogeography since it can explain the ubiquitous species-area constant $z \cong 1/4$ from $S = CA^z$, where S = species count, A = area and c and z are fitted constants (see Preston, 1962;

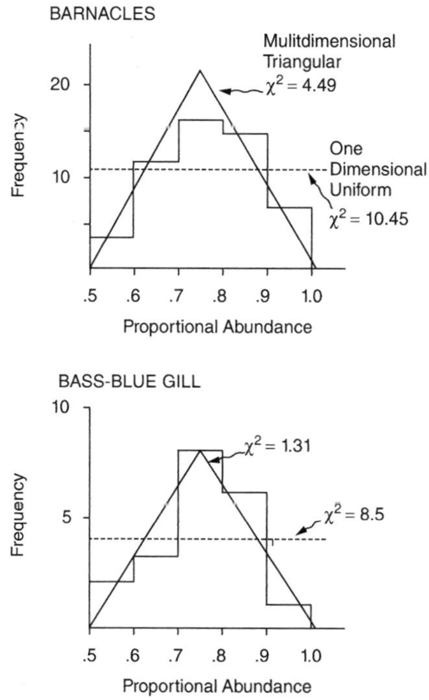

Figure 7.2 A test of one-dimensional breakage (MacArthur, 1957, 1960) versus multidimensional breakage for two-species assemblages of barnacles and fish. This figure shows that the relative abundance of most abundant species of fish (in biomass) and barnacles (in numbers of individuals) is not distributed uniformly but appears to cluster about the expected figure of 0.75. In both cases, the hypothesis of uniform breakage on one niche dimension is rejected at the 95% level.

MacArthur and Wilson, 1967; May, 1975; Connor and McCoy, 1979 for a catalogue of examples).

Under appropriate assumptions about the breakage distribution F (Figure 7.2) the niche hierarchy model can generate approximately the same relationship as equation 7.1 (Figure 7.3).[1] On the other hand, if one arbitrarily assumes that the successive breakages are uni-

[1]These assumptions are discussed in section 7.3.1 and Appendix D. It is worth noting that the result in Figure 7.3 will not be affected by the measure of abundance used as long as these measures scale as rough multiples of each other. For example, virtually the same standard deviation was obtained for the Hubbard Brook bird community (Holmes, 1978) regardless of whether numbers of individuals or biomass was used (2.57 vs 2.61).

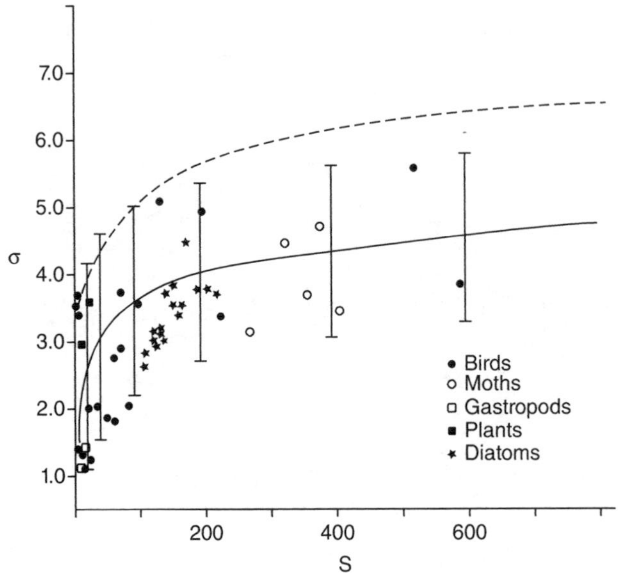

Figure 7.3 Graphical evidence for the sequential breakage hypothesis for an arbitrary assortment of taxonomic collections. The solid line and error bars represent the mean and two standard deviations of the mean predicted by the sequential breakage model. The dashed line represents the trend generated by biologically trivial or arbitrary sequential breakages, i.e., where the successive breakage fractions are chosen uniformly from the interval (0,1). This latter trend is identified with the null hypothesis that species are artifacts of classification in the nominalist sense.

form in the interval (0,1), then the canonical relationship is not obtained (dashed curve in Figure 7.3). This rules out an alternative and biologically trivial interpretation of breakage proposed by Aitchison and Brown (1967) where arbitrary hierarchical systems of classification could create sequential breakage. Clearly, then, the canonical lognormal distribution cannot be explained as an artifact of classification in the nominalist sense, and species do not represent arbitrary subdivisions of communal biomass.

These results demonstrate that a reasonable extension of the breakage rule derived from the two-species case can generate the ubiquitous canonical lognormal distribution, whereas arbitrary sequential breakages will not account for the relationship. Furthermore, because the model generates the canonical lognormal distribution, it can also explain the Preston-MacArthur-Wilson species-area constant $z \cong \frac{1}{4}$. Therefore insofar as an underlying functional hierarchy is reflected in the species abundance pattern one would expect to find these large-scale regularities in the abundance data. The next section examines this correspondence more carefully.

7.2 SUPPORTING EVIDENCE FOR NICHE HIERARCHY

7.2.1 The Geometric Series and Broken Stick Patterns as Special Cases

The geometric series and broken stick patterns are typically observed in small assemblages containing fewer than ten species (Pielou, 1975). It is argued that only small assemblages are sufficiently homogeneous for these patterns to obtain. Data fitted to the geometric series distribution are typically more equitable than predicted, and those fitting the broken stick are often less equitable than the model predicts.

It is interesting that both of these patterns are within the range of variation that could be accounted for by the niche hierarchy model. If one assumes conservatively that F is a split delta distribution with all its mass centered at 0.25 and 0.75, then the geometric series pattern may be obtained by non-randomly choosing the less abundant species to divide resources with at each step. Likewise, a distribution more even than the broken stick is attainable if the largest piece of resource pie is selectively subdivided at each stage (see Figure 7.4).

More significantly, the fact that these patterns are restricted to small ensembles would be predicted by the niche hierarchy model. That is, sequential breakages are most likely to produce these divergent abundance distributions in small ensembles, before the law of large numbers operates. In this sense, therefore, the geometric series and broken stick patterns may be considered special cases of sequential breakage.

7.2.2 A Deductive Approach

The preceding work assumed that an underlying functional hierarchy was responsible for defining the bifurcations that sequentially split up

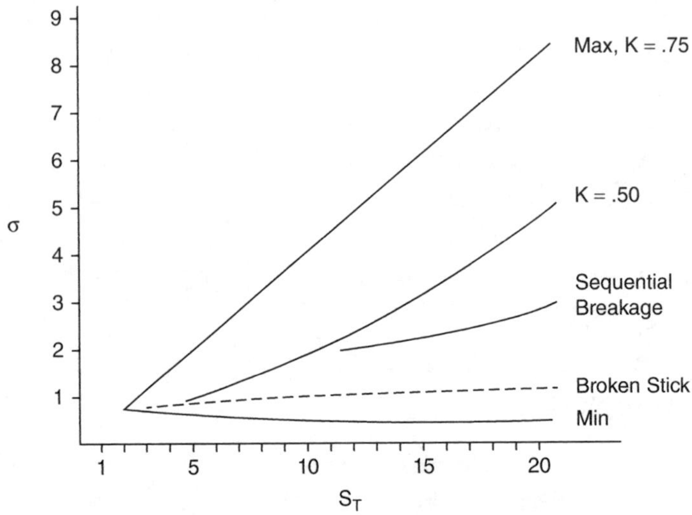

Figure 7.4 The geometric series and broken stick patterns as special cases of sequential breakage. The figure plots disequitability (measured by the standard deviation σ of logarithmic sizes) against species count S for the three models. Min and Max denote the range of abundance patterns encompassed by the niche hierarchy model and the central line denotes the expectation.

the communal resource pie. There, an inductive approach was taken to generate large-scale patterns by extrapolating the triangular rule observed in the two-species case. Our aim here will be to complement this inductive approach by seeing if it is possible to recover the extrapolated version of the triangular pattern from the larger multi-species situation. This involves dissecting a niche overlap dendrogram to see if it is possible to recover the breakage rule F from the abundances of a larger ensemble. If the model is correct, the deduced distribution should be capable of generating the canonical coupling between S and σ observed above.

Data used will be from the Hubbard Brook bird community study by Holmes, Bonney and Pacala (1979). This study was chosen because it contained a sufficiently large dendrogram together with the species abundance data required for this test.

To recover F, the abundance of each species (grams/10 ha.) is assigned to the appropriate terminal branch in the dendrogram.

Proportions are then computed for each bifurcation by summing abundances for all species contained in each subtree. The empirical distribution F is then rendered by plotting these proportions for all bifurcations. Results for the Hubbard Brook community are presented in Figure 7.5. The proportions plotted here include abundance data for five years (1969–1973). Distributions from any one year closely resemble the composite F where, as shown in the figure, values seem to cluster as in the two-species case. Table 7.1 shows that this F is roughly capable of generating the canonical coupling between S and σ.

These results corroborate the proposed correspondence between functional organization and species abundance patterns. By partitioning the ensemble along lines determined by the functional portrait we were able to recover the predicted breakage rule F.

To judge the significance of this result it is necessary to compare it with a random null hypothesis. It must be determined whether these patterns in fact depend on a particular relationship between abundance and functional organization or whether they could be generated at random. As a conservative null hypothesis we will see whether such results can be obtained when abundance is randomly related to

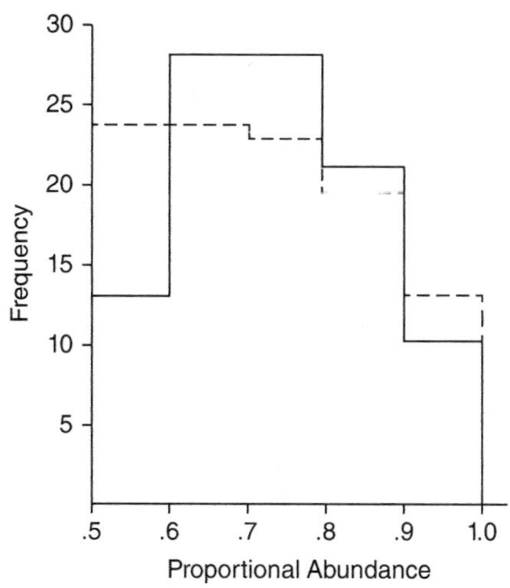

Figure 7.5 Deduced upper half of F for the Hubbard Brook community studied by Holmes et. al., (1979). Dashed line is the null hypothesis generated by randomly reshuffling abundances in the dendrogram.

Table 7.1 Test of Breakage Distribution F Deduced from Hubbard Brook Data (Holmes et. al., 1979)

Canonical		Deduced F	Random F
s	σ	$\tilde{E}(\sigma)$	$\tilde{E}(\sigma)$
25.	2.97	2.86	3.41
50.	3.37	3.29	3.74
100.	3.72	3.74	4.23
200.	4.05	4.17	4.61
400.	4.36	4.49	5.39

functional structure. This is a conservative null hypothesis in that the original set of abundances and the structure of the dendrogram will remain unaltered, only their relationship to each other will be randomized. This randomization simply involves an arbitrary reshuffling of abundances in the terminal branches of the dendrogram. If there is no special correspondence between abundance and organization then it should be possible to recreate the deduced patterns in Figure 7.5 and Table 7.1.

The breakage distribution F that results from this null hypothesis is indicated by the dashed line in Figure 7.5. Using a chi-square test, this randomly generated F is found to differ significantly ($p < 0.05$) from the deduced F.[2] Furthermore the randomly generated breakage rule is no longer able to produce the canonical relationship in Table 7.1. These preliminary results suggest that the functional portrait of the niche is tied non-randomly to species abundance. Partitioning the community along functional lines yields a predicted form of F, whereas randomly shuffling abundances does not. This constitutes an independent test of the niche hierarchy model.

[2]A stronger test can be achieved numerically by simulating the reshuffling process and performing a Kolmogorroff-Smirnov test on the cumulative frequencies to determine empirical confidence limits.

7.2.3 A Direct Approach

To determine more directly whether a correspondence exists between functional organization and abundance patterns, one can look at the relationship between the position that a species occupies in a dendrogram and its relative abundance. According to the niche hierarchy model a species whose niche is highly nested in a dendrogram will possess a corresponding abundance or "niche fragment" that has been subdivided many times. This niche fragment in turn should tend to be smaller than one involved in only a few subdivisions. It follows that one should expect to find, on average, an inverse relationship between the abundance of a species and its position or rank in a dendrogram.

To test this idea we will use Szaro and Balda's (1975) study of four small bird communities from a ponderosa pine forest. This interesting and thorough work examines community structure in habitats that were disturbed to varying degrees, ranging from a pristine control site to a severely thinned one. This was a rare study in that it contained the necessary dendrogram and abundance data required for the test. We will also consider the larger community from Hubbard Brook that was studied in the previous section. The ponderosa pine communities were not included there because they contained too few species to generate a reliable F.

Using the total weight of each species population (gms/40 ha. for the ponderosa pine communities, and gms/10 ha. for the Hubbard Brook community), Figure 7.6 plots abundance against species rank (number of bifurcations in the dendrogram) for each community.[3] As can be seen, the four ponderosa pine communities had negative correlations whereas the relationship for the Hubbard Brook community was not at all clear. Because a given dendrogram represents a single sample path (rendering) of a sequential breakage process a clear coupling will not be realized in every specific instance. Table 7.2 shows the t-values for each of these regressions when abundances are untransformed, log-transformed, and ranked. Although only the silviculturally cut and strip cut plots had significant t-values (5% level),

[3]The broad tailed hummingbird was omitted from the calculations as an anomalous species.

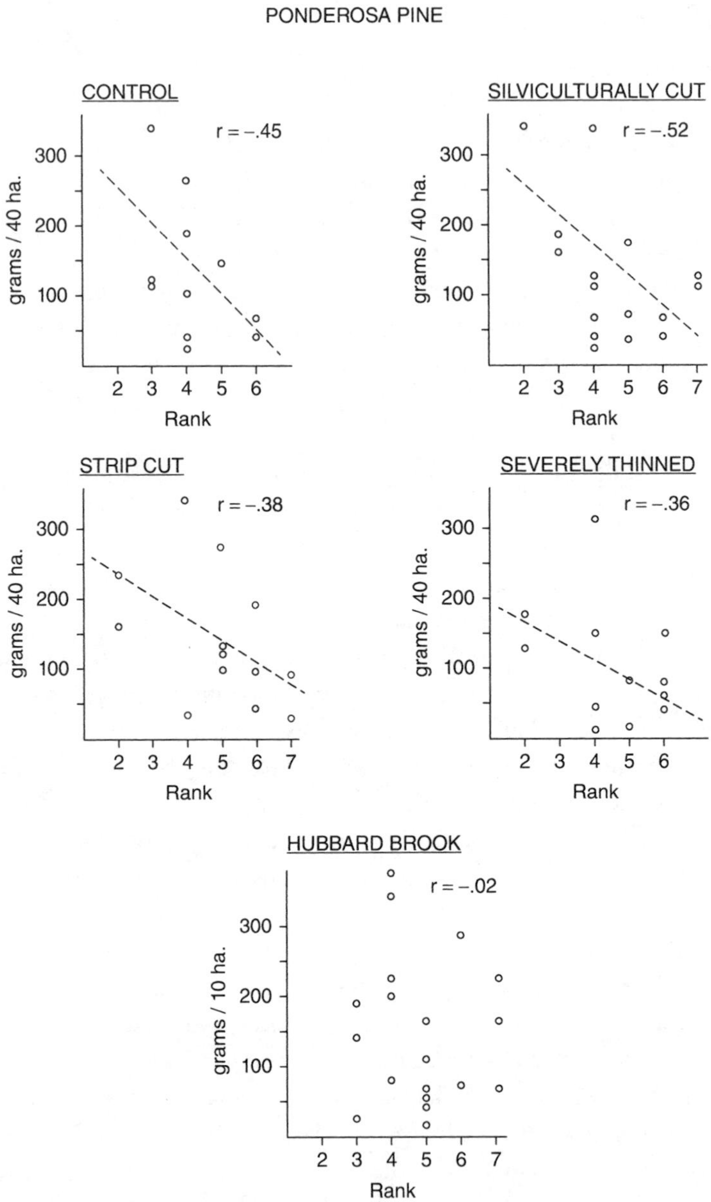

Figure 7.6 Abundance versus species rank (position in the dendrogram) for five bird communities (Szaro and Balda, 1975; Holmes et. al., 1979). The four ponderosa pine communities all show significant negative correlations as predicted by the niche hierarchy model. (See Table 7.2.)

Table 7.2 t-values for regressions of abundance against rank in the dendrogram for five bird communities.

	Control	Silviculturally cut	Strip cut	Severly thinned	Hubbard Brook
Untransformed (Figure 7.6)	−1.36	−1.92	−1.58	−0.82	−0.20
log-Rank	−1.31	−1.68	−1.74	−0.69	−0.06
Rank-Rank	−1.23	−0.97	−2.50	−1.28	0.14

one can make a stronger statement that a negative relationship exists between abundance and rank by looking at an ensemble statistic. In particular, for the untransformed and log-transformed regressions, all 5 examples had negative t-values. Because t is distributed symmetrically about zero the probability of this happening at random is 2^{-5}. Thus at the 3% level these data appear to support the idea that the functional portrait of the niche is closely tied to the species abundance pattern in a way predicted by the niche hierarchy model.

Among the ponderosa pine communities it is interesting that the inverse relationship was steepest and clearest for the relatively undisturbed habitats (control and silviculturally thinned) than for the more heavily disturbed ones (strip cut and severely thinned). Because the weight per species was roughly the same in each habitat (130 gms. control; 145 gms. silv cut; 133 gms. strip cut; 103 gms. severely thinned) this suggests the possibility of an equilibrium requirement for the pattern to obtain.

7.2.4 Implications of a Correspondence Between Organization and Abundance

As suggested in section 7.2.1, the geometric series and broken stick distributions can be roughly thought of as extremes in a continuum of possible abundance patterns. If our hypothesis is correct and abundances do reflect hierarchical organization in the niche, these extremes in abundance patterns should be matched by extremes in the morphology of functional hierarchies.

Consider the two dendrograms in Figure 7.7. Structure A represents a simple linear hierarchy of niche similarity while community

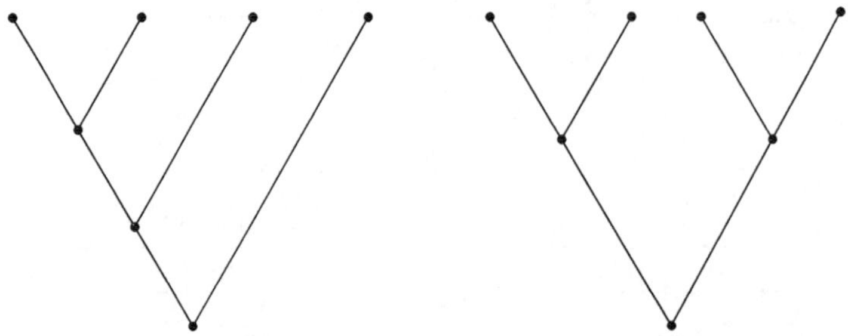

Figure 7.7 Two extreme functional portraits. According to the niche hierarchy model A should correspond to a less equitable geometric series-like abundance distribution and model B should identify with a more equitable one.

B is evenly divided. If, as demonstrated above, the final abundance of a species is inversely proportional to the number of niche subdivisions, then structure A should correspond to an uneven geometric series-like distribution and B should identify with a more equitable abundance pattern. That is, extremes in the shapes of the functional portraits should correspond to extremes in the relative abundance distributions.

A simple index which can be used to measure the shape of a dendrogram is suggested by the following identity for random binary trees (Appendix D):

$$\rho = E(N_1 + N_1 N_2 - N_1^2) = 1 \tag{7.2}$$

where N_1 is the distance from terminal vertex v_1 to the nearest common ancestor of v_1 and v_2 (see Figure 7.8). This theorem states that for random binary trees with S endpoints, the expected distance between any two endpoints from a common ancestor may be combined to form cross-moments that generate the above simple identity, $\rho = 1$. It is easy to see that for a perfectly uneven geometric series structure,

$$\rho_A = \frac{2 + \sum_{k=2}^{S-1} (s + 1 - k)(3k - k^2)}{\binom{S}{2}} \tag{7.3}$$

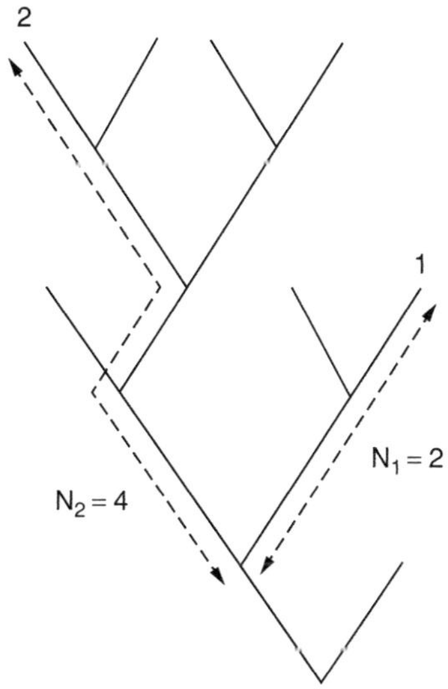

Figure 7.8 A random binary tree.

and for the perfectly even case,

$$\rho_B = \frac{\sum_{k=1}^{S-1} 2^{k+1} k}{\binom{S}{2}} \tag{7.4}$$

Thus $\rho_A < 1 < \rho_B$ for $S > 2$. Hence one can characterize the structure of a dendrogram by ρ, where $\rho > 1$ indicates an even structure and $\rho < 1$ denotes an uneven structure. Insofar as abundances correspond to the organization of a community, ρ represents the functional equivalent of a species diversity index. It would be useful at some future date to derive an expression for the variance of ρ so that a test could be constructed to measure the significance that a given tree deviates from random.

An interesting alternative test of the correspondence between organization and abundance could be accomplished by comparing ρ with some measure of equitability in the species abundance distribution. Unfortunately, the communities from the two studies examined above have a very similar functional organization and almost identical equitabilities. It would be worthwhile to check this correspondence at some future date once data for a full range of functional structures and abundance patterns becomes available.

In a thought provoking paper Harmsen (in press) has recently remarked that lepidopteran collections from pristine areas often tend to be more equitable than the canonical relationship predicts. These exceptionally diverse ensembles (ca. 500 to 700 species from Perth Road, Ontario) are argued to be older and less tampered with than the majority of other collections that have been seen to fit Preston's canonical lognormal distribution. It is suggested that a long evolutionary history combined with a minimum of disturbance has allowed these communities to evolve a "non-minimal" or non-random community structure.

In terms of the functional portrait such non-minimal structure would be expressed as an increase in ρ; $\rho \rangle 1$. This may be expected if an environment becomes saturated with species and the rare ones become crowded out by failing to meet a minimum population size (Mayr, 1970). This threshold effect would tend to place a limit on the highly bifurcated lines, and favor niche sudivisions (e.g., by speciation or colonization) along the less heavily carved branches of the dendrogram. In the end, this process would produce a more evenly structured functional portrait and greater equitability in the abundances. Thus threshold culling of rare species may cause older and more stable communities in species-saturated environments to deviate from the random structure associated with the canonical hypothesis. This agrees with several observations on moths (Hebert, 1979) and birds (MacArthur, 1969; Preston, 1982) where species-rich tropical collections are more equitable than their canonical temperate counterparts.

7.3 RELATED RESULTS

In closing, we briefly mention two other findings that pertain to the niche hierarchy model.

7.3.1 Moments of Species Abundance Distributions

The original results presented for the canonical coupling between species count and variance were obtained by Monte Carlo methods.

An analytical solution for this relationship has now been found (Siegel and Sugihara, 1982) for the expected variance as a function of the number of breakages involved (S – 1). This solution utilizes the following auxiliary moments of the breakage distribution F which is assumed to be symmetric on (0,1),

$$\mu_m = E(\log(W))^m$$
$$\eta = E(\log(W)\ \log(1 - W))$$

where W is an observation from F. The expected sample variance for S species is:

$$E(\sigma_s^2) = \{2(1 + \frac{2}{S-1})\sum_{k=2}^{S}\frac{1}{k} - 2\}\ \mu_2 - \eta \tag{7.5}$$

The proof of this theorem is contained in Appendix D.

The important thing to notice in equation 7.5 is that the same coupling will be obtained for any distributions F on (0,1) having the same second moment. Thus a wide variety of assumptions on F will generate the canonical relationship.

7.3.2 Abundance Patterns Implied by Dynamic Models

As a preliminary attempt to find an explanation, other than multidimensional apportionment, for the triangular F observed in the two-species case, I have computed distributions of abundance for the equilibrium solutions of the Gilpin-Ayala (1973) generalized version of the Lotka-Volterra equations,

$$\frac{dN_1}{dt} = \eta_1 N_1 \left\{K_1 - \left[\frac{N_1}{K_1}\right]^\theta - \alpha_{12}\left[\frac{N_2}{K_2}\right]^\theta\right\}$$

where θ is a free parameter that can be tuned to fit a variety of isoclines of different curvature. For equal K values, and α's distributed uniformly on the interval (0,1), the distributions of abundance expected are:

$$\text{if}\quad \theta = 1,\quad g(y) = y^{-2}$$
$$\theta = 2,\quad g(y) = 2(1 - y)y^{-3}$$
$$\theta = \frac{1}{2},\quad g(y) = \frac{1}{2}(1 - y)^{-1/2}y^{-3/2}$$

None of these densities come close to the triangular pattern observed in the two-species case. These results are qualitatively robust for a wide variety of assumptions on the distributions of K and α (see Sugihara, 1981 for method of analysis).

SUMMARY

This chapter presents evidence for a correspondence between two classical ideas about structure in ecological communities: functional organization and species abundance patterns. Such correspondence is suggested in a recently proposed species abundance hypothesis, the niche hierarchy model. Here it is argued that the functional portrait given in a niche overlap dendrogram, implies that abundances in a community should be divided up sequentially. When these sequential divisions involve multiple niche factors the expected large-scale abundance pattern generated is the ubiquitous canonical lognormal distribution (Preston, 1962). Several independent tests of the model are discussed to demonstrate that species abundances are predictably and non-randomly tied to functional organization in communities.

APPENDIX A

Summary of Corrections for Briand (1982) Webs:

Web I.D.	Correction	Reason
6	add 1 to (25, 18)	miscopied from original article
6	add 1 to (23, 19)	scoter eats flounder (23, 9) and clams
8	add 1 to (11, 6)	striped mummichog eats amphipods
18	delete 1 from (22, 27)	skinks do not eat gekkos
20	add 1 to (10, 4)	predatory planktonic invertebrates (ctenophora, sagitta etc.) eat copepods
24	add 1 to (4, 8)	maryland yellow throat, yellow warbler, song sparrow eat insects
33	add 1 to (10, 33)	barbus paludinosa eats terrestial
	add 1 to (10, 19)	insects and cladocera (Pienaar 1968)
40	add 1 to (4, 4)	predatory insects eat other insects

APPENDIX B

Niche Overlap Graph $G_R(C;\lambda)$, Resource Graph $G_C(R;\lambda^{-1})$ and Food Web Matrix for Corrected Briand Communities

CONSUMER OVERLAP $G_R(C;\lambda)$

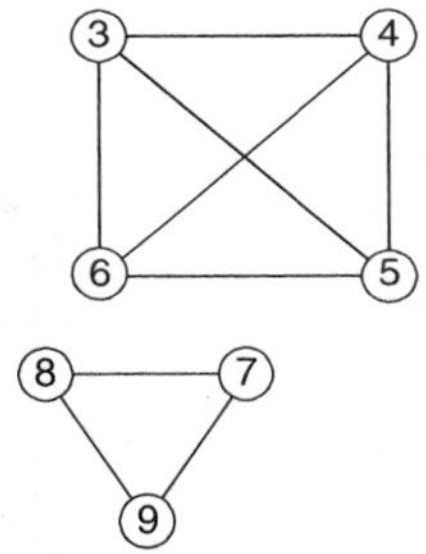

RESOURCE GRAPH $G_C(R;\lambda^{-1})$

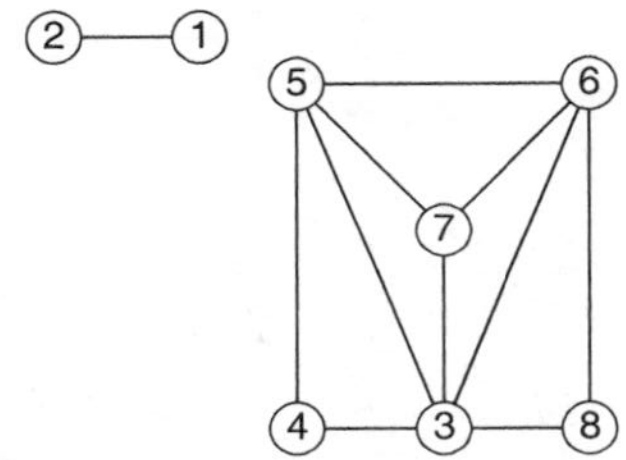

WEB 1: COCHIN BACKWATER

1 – AUTOTROPHS
2 – DETRITUS
3 – SHRIMPS
4 – BENTHOS
5 – ZOOPLANKTON HERBIVORES
6 – FISH HERBIVORES
7 – OTHER AQUATIC CARNIVORES
8 – FISH CARNIVORES
9 – MAN

FOOD WEB MATRIX

	3	4	5	6	7	8	9
1	1	1	1	1	0	0	0
2	1	1	1	1	0	0	0
3	0	0	0	0	1	1	1
4	0	0	0	0	1	0	0
5	0	0	0	0	1	1	0
6	0	0	0	0	0	1	1
7	0	0	0	0	0	1	0
8	0	0	0	0	0	0	1

CONSUMER OVERLAP $G_R(C;\lambda)$

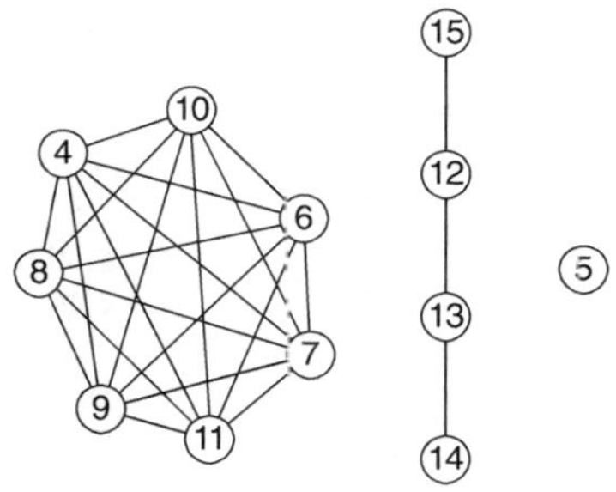

RESOURCE GRAPH $G_C(R;\lambda^{-1})$

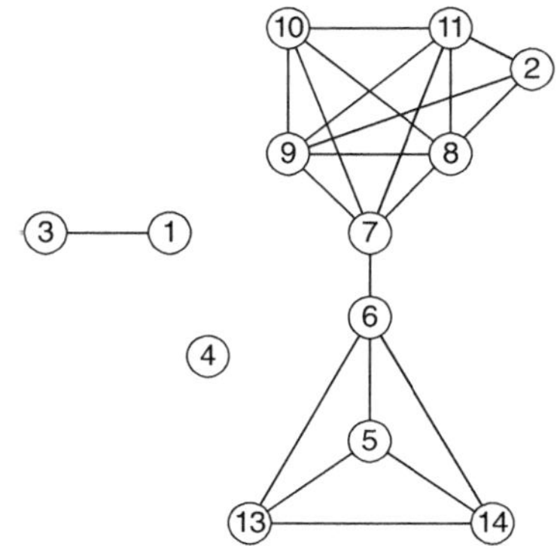

WEB 2: KNYSNA ESTUARY

1 – PHYTOPLANKTON
2 – ATTACHED PLANTS
3 – ORGANIC DETRITUS
4 – ZOOPLANKTON
5 – HYPROHAMPHUS
6 – MUGIL
7 – LIPOGEBIA
8 – LAMYA
9 – SOLEN
10 – ARENICOLA
11 – HYMENOSOMA
12 – JOHNIUS
13 – LITHOGNATHUS
14 – RHABDOSARGUS
15 – HYPACANTHUS

FOOD WEB MATRIX

	4	5	6	7	8	9	10	11	12	13	14	15
1	1	0	0	0	1	1	0	0	0	0	0	0
2	0	0	0	0	0	0	0	0	0	0	1	0
3	1	0	1	1	1	1	1	1	0	0	0	0
4	0	1	0	0	0	0	0	0	0	0	0	0
5	0	0	0	0	0	0	0	0	0	0	0	1
6	0	0	0	0	0	0	0	0	1	0	0	1
7	0	0	0	0	0	0	0	0	1	1	0	0
8	0	0	0	0	0	0	0	0	0	1	1	0
9	0	0	0	0	0	0	0	0	0	1	1	0
10	0	0	0	0	0	0	0	0	0	1	0	0
11	0	0	0	0	0	0	0	0	0	1	1	0
13	0	0	0	0	0	0	0	0	0	0	0	1
14	0	0	0	0	0	0	0	0	0	0	0	1

CONSUMER OVERLAP $G_R(C;\lambda)$

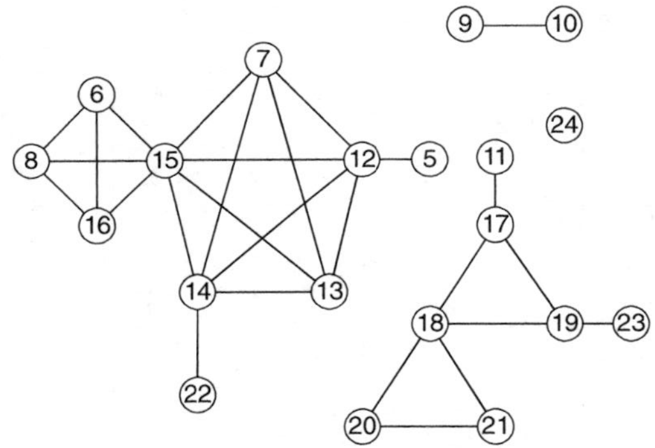

RESOURCE GRAPH $G_C(R;\lambda^{-1})$

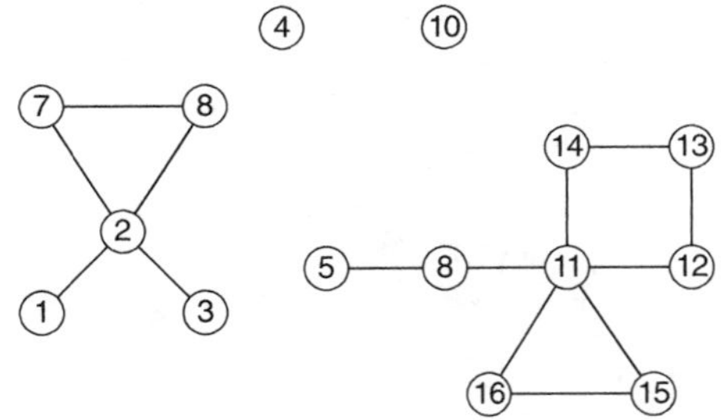

WEB 3: LONG ISLAND ESTUARY

1 – ORGANIC DEBRIS
2 – WATER PLANT
3 – PLANKTON
4 – MARSH PLANTS
5 – BAY SHRIMP
6 – SILVERSIDES
7 – MUD SNAIL
8 – CLAM
9 – MOSQUITO
10 – CRICKET
11 – BILLFISH
12 – EEL
13 – FLUKE
14 – BLOWFISH
15 – MINNOW
16 – MINNOW
17 – TERN
18 – OSPERY
19 – GREEN HERON
20 – MERGANSER
21 – CORMORANT
22 – GULL
23 – KINGFISHER
24 – REDWING BLACKBIRD

FOOD WEB MATRIX

	5	6	7	8	9	10	11	12	13	14	15	16	17	18	19	20	21	22	23	24
1	1	0	0	0	0	0	0	1	0	0	0	0	0	0	0	0	0	0	0	0
2	0	0	1	0	0	0	0	1	1	1	1	0	0	0	0	0	0	0	0	0
3	0	1	0	1	0	0	0	0	0	0	1	1	0	0	0	0	0	0	0	0
4	0	0	0	0	1	1	0	0	0	0	0	0	0	0	0	0	0	0	0	0
5	0	0	0	0	0	0	1	0	0	0	0	0	0	0	0	0	0	0	0	0
6	0	0	0	0	0	0	1	0	0	0	0	0	1	0	0	0	0	0	0	0
7	0	0	0	0	0	0	0	0	0	1	0	0	0	0	0	0	0	0	0	0
8	0	0	0	0	0	0	0	0	0	1	0	0	0	0	0	0	0	1	0	0
10	0	0	0	0	0	0	0	0	0	0	0	0	0	0	0	0	0	0	0	1
11	0	0	0	0	0	0	0	0	0	0	0	0	1	1	1	0	0	0	0	0
12	0	0	0	0	0	0	0	0	0	0	0	0	0	1	0	1	1	0	0	0
13	0	0	0	0	0	0	0	0	0	0	0	0	0	0	0	1	1	0	0	0
14	0	0	0	0	0	0	0	0	0	0	0	0	0	1	0	1	0	0	0	0
15	0	0	0	0	0	0	0	0	0	0	0	0	0	0	1	0	0	0	1	0
16	0	0	0	0	0	0	0	0	0	0	0	0	0	0	1	0	0	0	1	0

CONSUMER OVERLAP $G_R(C;\lambda)$

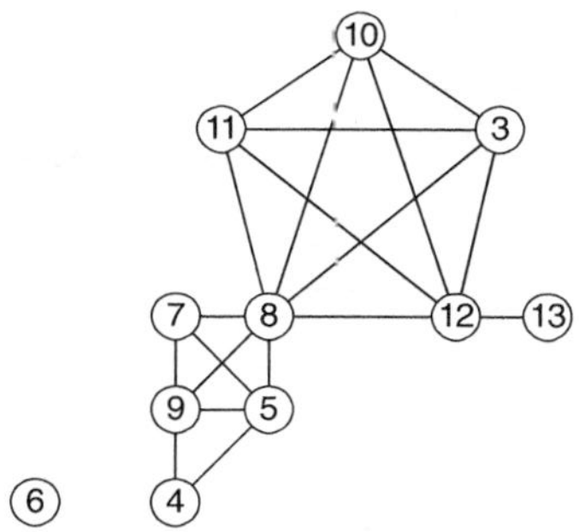

RESOURCE GRAPH $G_C(R;\lambda^{-1})$

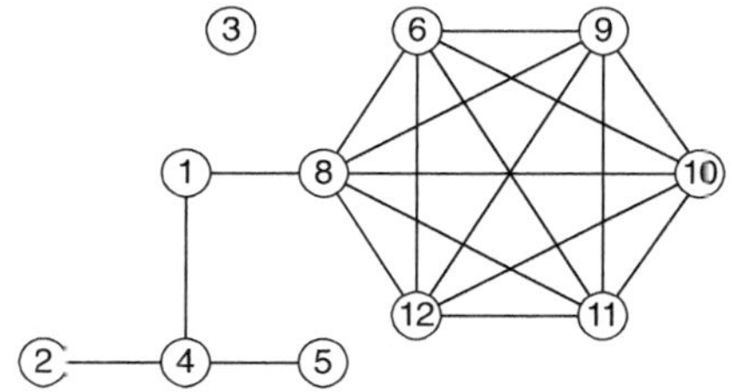

WEB 4: SALICORNIA SALT MARSH

1 – TERRESTRIAL PLANTS
2 – MARINE PLANTS
3 – TERRESTRIAL INVERTEBRATES
4 – INTERTIDAL, MARINE INVERTEBRATES
5 – FISHES
6 – SOREX
7 – HERONS
8 – RALLUS AMAS
9 – MIGRANT SHORE BIRDS, WATERFOWL
10 – PASSERINES
11 – MICROTUS REITHRODONTOMYS
12 – RATTUS
13 – CIRCUS AS10

FOOD WEB MATRIX

	3	4	5	6	7	8	9	10	11	12	13
1	1	0	0	0	0	1	0	1	1	1	0
2	0	1	1	0	0	0	1	0	0	0	0
3	0	0	0	1	0	0	0	0	0	0	0
4	0	0	1	0	1	1	1	0	0	0	0
5	0	0	0	0	1	0	0	0	0	0	0
6	0	0	0	0	0	0	0	0	0	0	1
8	0	0	0	0	0	0	0	0	0	0	1
9	0	0	0	0	0	0	0	0	0	1	1
10	0	0	0	0	0	0	0	0	0	0	1
11	0	0	0	0	0	0	0	0	0	0	1
12	0	0	0	0	0	0	0	0	0	0	1

CONSUMER OVERLAP $G_R(C;\lambda)$

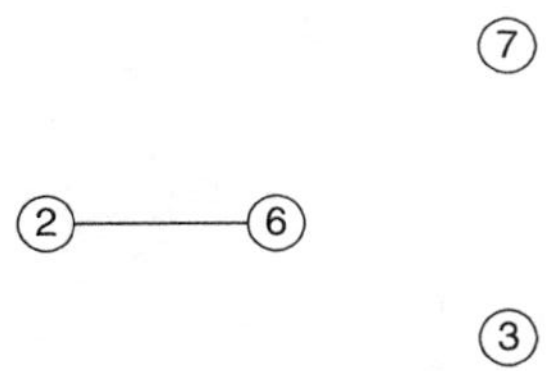

RESOURCE GRAPH $G_C(R;\lambda^{-1})$

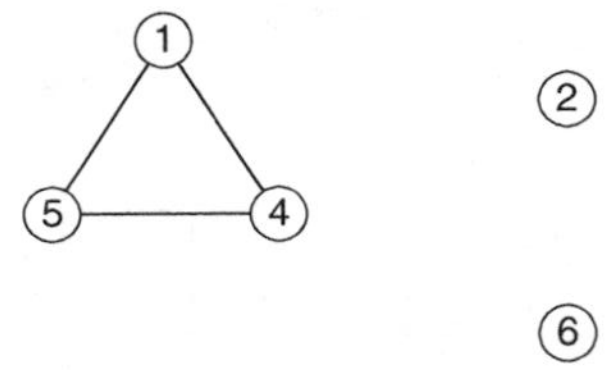

WEB 5: SALT MARSH, GEORGIA

1 – SPARTINA
2 – HERBIVOROUS INSECTS
3 – SPIDERS, PASSERINES, DRAGONFLIES
4 – ALGAE
5 – BACTERIA
6 – UCA, SESARNA, MODIOLUS, LITTORINA, OLIGOCHAETE, ETC.
7 – EURYTHIUEM, CLAPPER RAIL, RACCOON

FOOD WEB MATRIX

	2	3	6	7
1	1	0	1	0
2	0	1	0	0
4	0	0	1	0
5	0	0	1	0
6	0	0	0	1

CONSUMER OVERLAP $G_R(C;\lambda)$

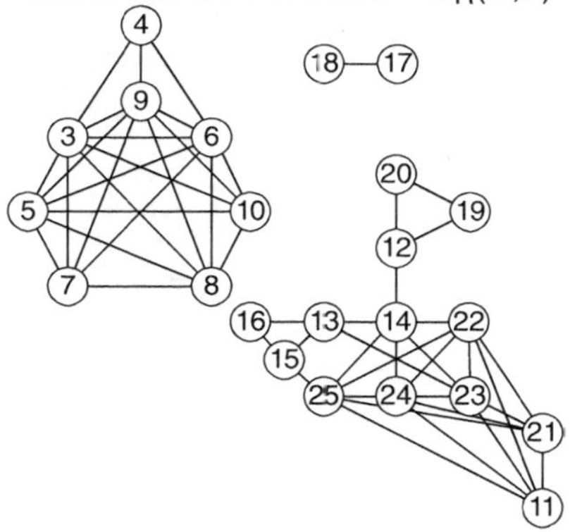

RESOURCE GRAPH $G_C(R;\lambda^{-1})$

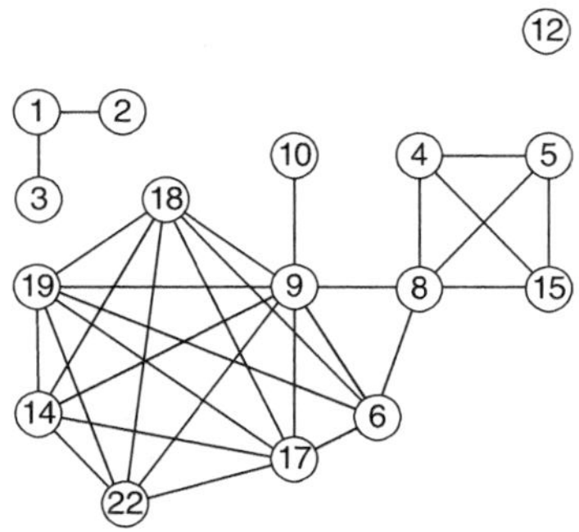

WEB 6: CALIFORNIA TIDAL FLAT

1 – PRIMARY PRODUCERS
2 – DETRITUS
3 – PROTOZOA
4 – CRUSTACEA
5 – CALLIANASSA
6 – SHORE CRABS
7 – PHORONOPSIS
8 – URECHIS
9 – CLAMS
10 – LUMBRINEREIS, NOTOMASTUS
11 – HERON
12 – CLEVELANDIA
13 – WADING BIRDS
14 – STINGRAY
15 – NEREIS, GLYCERA
16 – NEMERTEANS
17 – TERNS, SHOREBIRD
18 – CABEZON
19 – FLOUNDER
20 – CEREBRATULUS
21 – LOON, CORMORANT
22 – STRIPED BASS
23 – SCOTER
24 – GULLS
25 – MAN

FOOD WEB MATRIX

	3	4	5	6	7	8	9	10	11	12	13	14	15	16	17	18	19	20	21	22	23	24	25
1	1	1	0	1	0	0	1	0	0	0	0	0	0	0	0	0	0	0	0	0	0	0	0
2	1	0	1	1	1	1	1	1	0	0	0	0	0	0	0	0	0	0	0	0	0	0	0
3	0	1	0	0	0	0	0	0	0	0	0	0	0	0	0	0	0	0	0	0	0	0	0
4	0	0	0	0	0	0	0	0	0	1	0	0	0	0	0	0	0	0	0	0	0	0	0
5	0	0	0	0	0	0	0	0	0	1	0	0	0	0	0	0	0	0	0	0	0	0	0
6	0	0	0	0	0	0	0	0	0	0	0	1	0	0	0	0	0	0	0	1	1	1	0
8	0	0	0	0	0	0	0	0	0	1	0	1	0	0	0	0	0	0	0	0	0	0	0
9	0	0	0	0	0	0	0	0	0	0	1	1	0	0	0	0	0	0	0	0	1	1	1
10	0	0	0	0	0	0	0	0	0	0	1	0	1	1	0	0	0	0	0	0	0	0	0
12	0	0	0	0	0	0	0	0	0	0	0	0	0	0	1	1	0	0	0	0	0	0	0
14	0	0	0	0	0	0	0	0	0	0	0	0	0	0	0	0	0	0	0	0	0	0	1
15	0	0	0	0	0	0	0	0	0	1	0	0	0	0	0	0	1	1	0	0	0	0	0
17	0	0	0	0	0	0	0	0	0	0	0	0	0	0	0	0	0	0	0	0	0	1	1
18	0	0	0	0	0	0	0	0	1	0	0	0	0	0	0	0	0	0	1	1	1	0	1
19	0	0	0	0	0	0	0	0	1	0	0	0	0	0	0	0	0	0	1	0	1	1	1
22	0	0	0	0	0	0	0	0	0	0	0	0	0	0	0	0	0	0	0	0	0	0	1

CONSUMER OVERLAP $G_R(C;\lambda)$

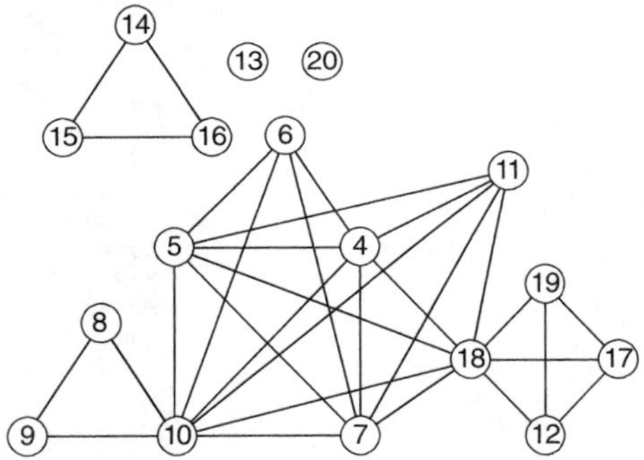

RESOURCE GRAPH $G_C(R;\lambda^{-1})$

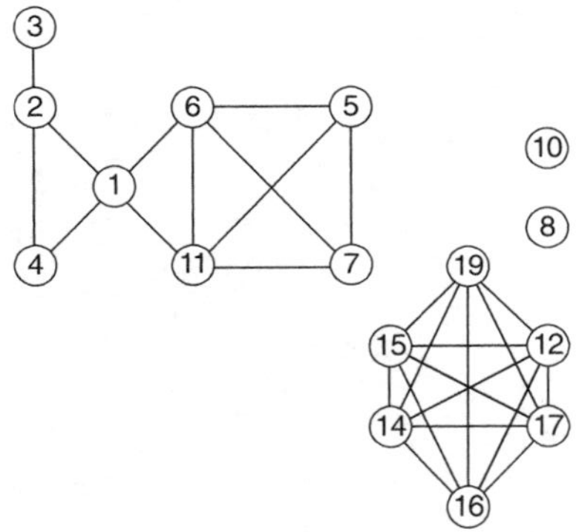

WEB 7: NARRAGANSET BAY

1 – FLAGELLATES, DIATOMS
2 – DETRITUS
3 – MACROALGAE, EELGRASS
4 – ACARTIA, OTHER COPEPODS
5 – SPONGES
6 – BENTHIC MACROFAUNA AND INFAUNA
7 – CLANS
8 – CTENOPHORES
9 – MEROPLANKTON, FISH LARVAE
10 – PACIFIC MENHADEN
11 – BIVALVES
12 – CRABS, LOBSTERS
13 – BUTTERFISH
14 – STRIPED BASS
15 – BLUEFISH
16 – MACKEREL
17 – OTHER DEMERSAL SPECIES
18 – STARFISH
19 – FLOUNDER
20 – MAN

FOOD WEB MATRIX

	4	5	6	7	8	9	10	11	12	13	14	15	16	17	18	19	20
1	1	1	0	1	0	0	1	1	0	0	0	0	0	0	1	0	0
2	1	1	1	1	0	0	1	0	0	0	0	0	0	0	0	0	0
3	0	0	1	0	0	0	0	0	0	0	0	0	0	0	0	0	0
4	0	0	0	0	1	1	1	0	0	0	0	0	0	0	0	0	0
5	0	0	0	0	0	0	0	0	0	0	0	0	0	0	0	1	0
6	0	0	0	0	0	0	0	0	1	0	0	0	0	1	1	1	0
7	0	0	0	0	0	0	0	0	0	0	0	0	0	0	0	1	0
8	0	0	0	0	0	0	0	0	0	1	0	0	0	0	0	0	0
10	0	0	0	0	0	0	0	0	0	0	1	1	1	0	0	0	0
11	0	0	0	0	0	0	0	0	0	0	0	0	0	1	1	1	0
12	0	0	0	0	0	0	0	0	0	0	0	0	0	0	0	0	1
14	0	0	0	0	0	0	0	0	0	0	0	0	0	0	0	0	1
15	0	0	0	0	0	0	0	0	0	0	0	0	0	0	0	0	1
16	0	0	0	0	0	0	0	0	0	0	0	0	0	0	0	0	1
17	0	0	0	0	0	0	0	0	0	0	0	0	0	0	0	0	1
19	0	0	0	0	0	0	0	0	0	0	0	0	0	0	0	0	1

CONSUMER OVERLAP $G_R(C;\lambda)$

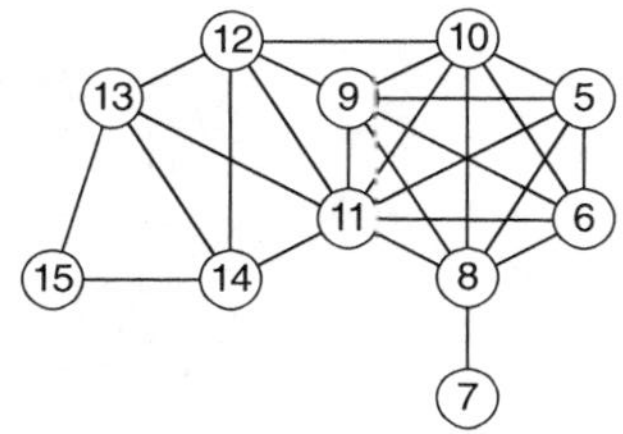

RESOURCE GRAPH $G_C(R;\lambda^{-1})$

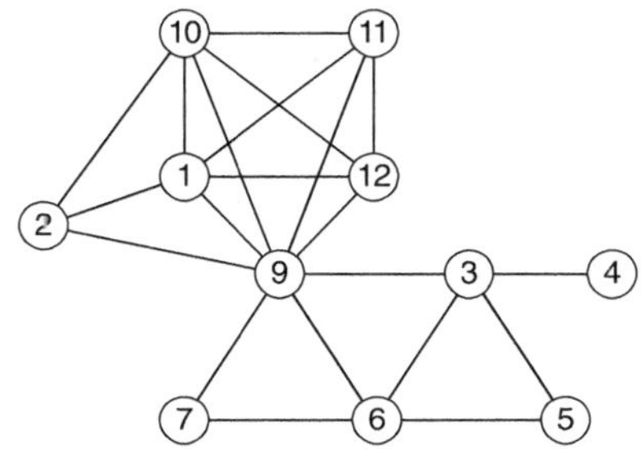

WEB 8: BISSEL COVE MARSH

1 – RUPPIA
2 – ULVA, ENTEROMORPHA
3 – DETRITUS
4 – PHYTOPLANKTON
5 – NEMATODES, COPEPODS, OSTRACODS
6 – AMPHIPODS
7 – ZOOPLANKTON
8 – JUVENILE MENHADEN
9 – SHRIMPS
10 – COMMON MUMMICHOG
11 – STRIPED MUMMICHOG
12 – OTHER FISH
13 – DUCKS
14 – EEL
15 – MAN

FOOD WEB MATRIX

	5	6	7	8	9	10	11	12	13	14	15
1	0	0	0	0	0	0	0	0	1	1	0
2	0	0	0	0	0	0	0	0	1	0	0
3	1	1	0	1	1	1	1	0	0	0	0
4	0	0	1	1	0	0	0	0	0	0	0
5	0	0	0	0	0	1	0	0	0	0	0
6	0	0	0	0	1	1	1	1	0	0	0
7	0	0	0	0	0	0	0	1	0	0	0
9	0	0	0	0	0	0	1	1	1	1	0
10	0	0	0	0	0	0	0	0	1	1	1
11	0	0	0	0	0	0	0	0	0	1	0
12	0	0	0	0	0	0	0	0	0	1	0

CONSUMER OVERLAP $G_R(C;\lambda)$

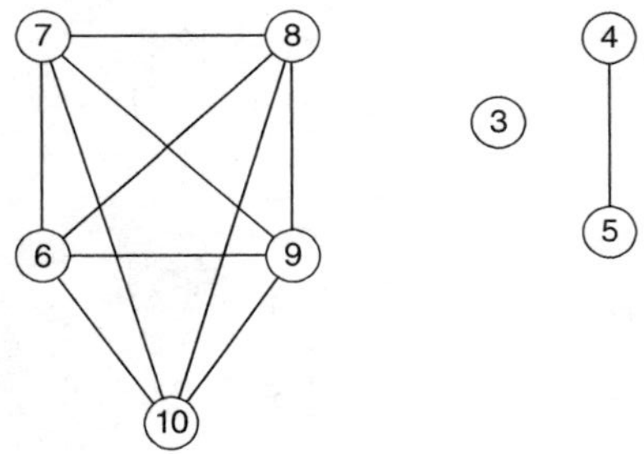

RESOURCE GRAPH $G_C(R;\lambda^{-1})$

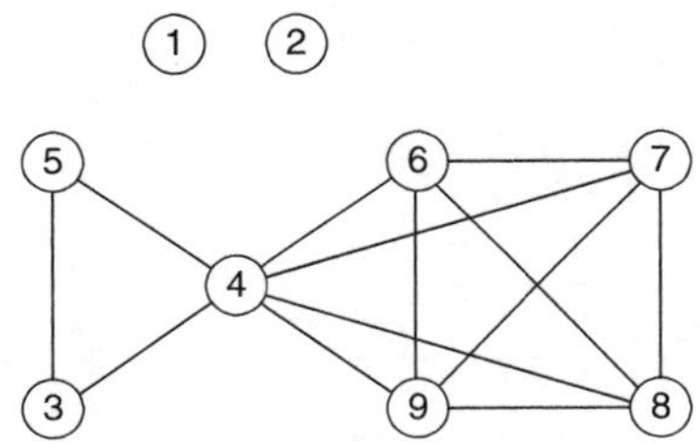

WEB 9: LOUGH INE RAPIDS

1 – ATTACHED ALGAE
2 – PLANKTON
3 – ANOMIA
4 – PARACENTROTUS
5 – GIBBULA CINERARIA
6 – CANCER
7 – PORTUNUS PUBER
8 – CARCINUS
9 – MARTHASTERIAS
10 – BIRDS

FOOD WEB MATRIX

	3	4	5	6	7	8	9	10
1	0	1	1	0	0	0	0	0
2	1	0	0	0	0	0	0	0
3	0	0	0	0	0	0	1	0
4	0	0	0	1	1	1	1	1
5	0	0	0	1	1	0	1	0
6	0	0	0	0	0	0	0	1
7	0	0	0	0	0	0	0	1
8	0	0	0	0	0	0	0	1
9	0	0	0	0	0	0	0	1

CONSUMER OVERLAP $G_R(C;\lambda)$

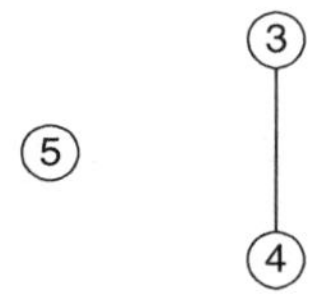

RESOURCE GRAPH $G_C(R;\lambda^{-1})$

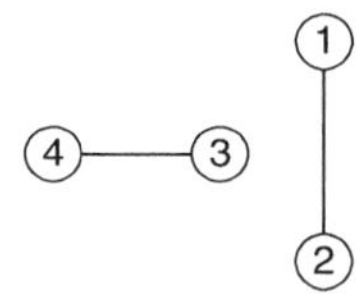

WEB 10: EXPOSED INTERTIDAL

1 – DETRITUS
2 – PLANKTON
3 – BALANUS BALANOIDES
4 – MYTILUS EDULIS
5 – THAIS LAPILLUS

FOOD WEB MATRIX

	3	4	5
1	1	1	0
2	1	1	0
3	0	0	1
4	0	0	1

CONSUMER OVERLAP $G_R(C;\lambda)$

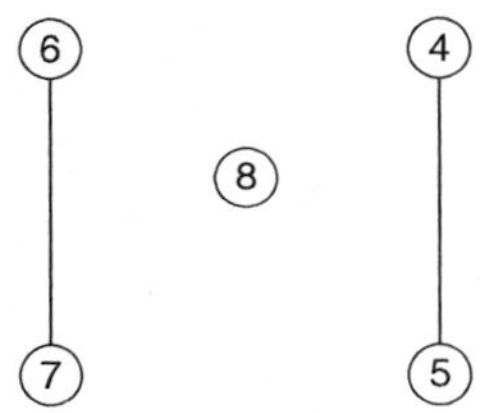

RESOURCE GRAPH $G_C(R;\lambda^{-1})$

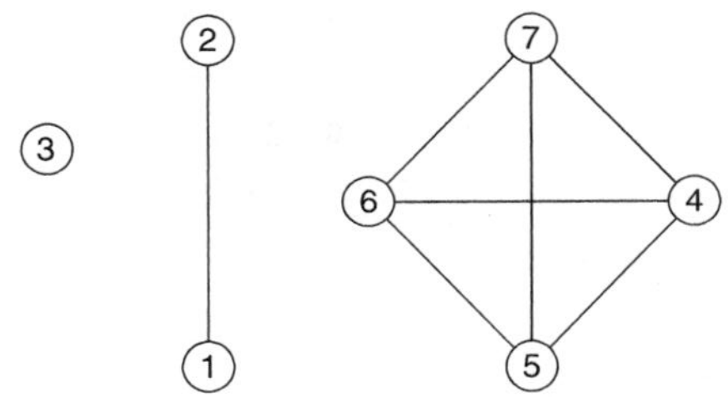

WEB 11: PROTECTED INTERTIDAL

1 – DETRITUS
2 – PLANKTON
3 – ALGAE
4 – ACMAEA FESTUDINALIS
5 – LITTORINA
6 – MYTILUS EDULIS
7 – BALANUS BALANOIDES
8 – THAIS LAPILLUS

FOOD WEB MATRIX

	4	5	6	7	8
1	0	0	1	1	0
2	0	0	1	1	0
3	1	1	0	0	0
4	0	0	0	0	1
5	0	0	0	0	1
6	0	0	0	0	1
7	0	0	0	0	1

CONSUMER OVERLAP $G_R(C;\lambda)$

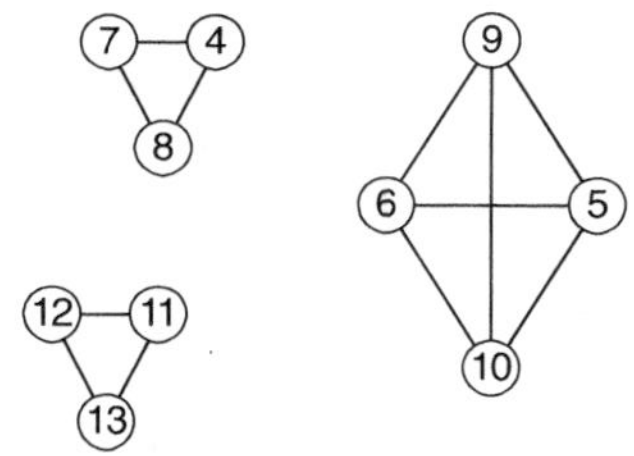

RESOURCE GRAPH $G_C(R;\lambda^{-1})$

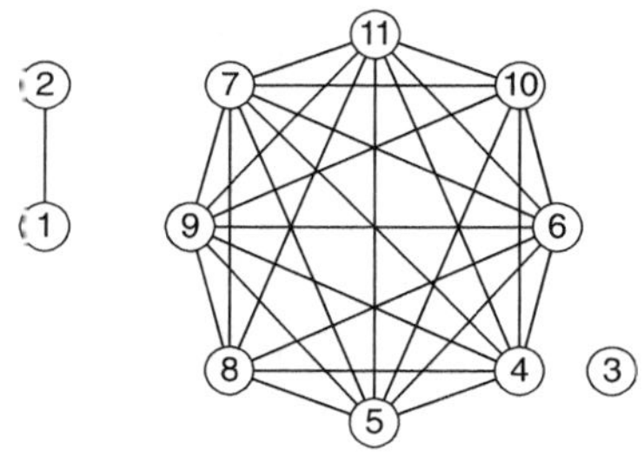

WEB 12: EXPOSED INTERTIDAL

1 – DETRITUS
2 – PLANKTON
3 – ALGAE
4 – ACORN BARNACLES
5 – LIMPETS
6 – TEGULA
7 – MYTILUS CALIFORNIANUS
8 – POLLICIPES
9 – CHITONS
10 – LITTORINA
11 – THAIS
12 – PISASTER
13 – LEPTASTERIAS

FOOD WEB MATRIX

	4	5	6	7	8	9	10	11	12	13
1	1	0	0	1	1	0	0	0	0	0
2	1	0	0	1	1	0	0	0	0	0
3	0	1	1	0	0	1	1	0	0	0
4	0	0	0	0	0	0	0	1	1	1
5	0	0	0	0	0	0	0	0	1	1
6	0	0	0	0	0	0	0	0	1	1
7	0	0	0	0	0	0	0	1	1	1
8	0	0	0	0	0	0	0	0	1	0
9	0	0	0	0	0	0	0	0	1	1
10	0	0	0	0	0	0	0	0	0	1
11	0	0	0	0	0	0	0	0	1	1

CONSUMER OVERLAP $G_R(C;\lambda)$

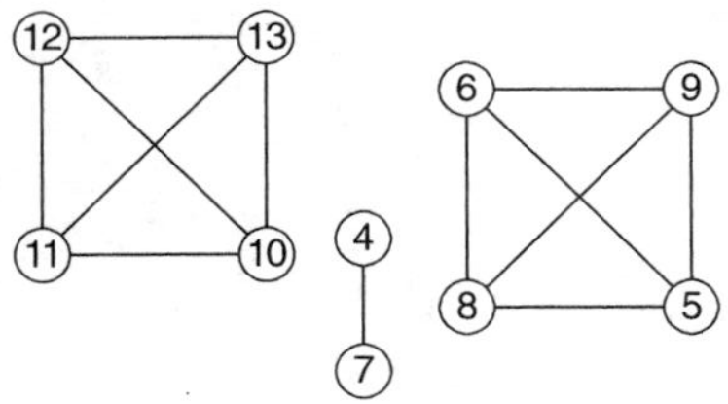

RESOURCE GRAPH $G_C(R;\lambda^{-1})$

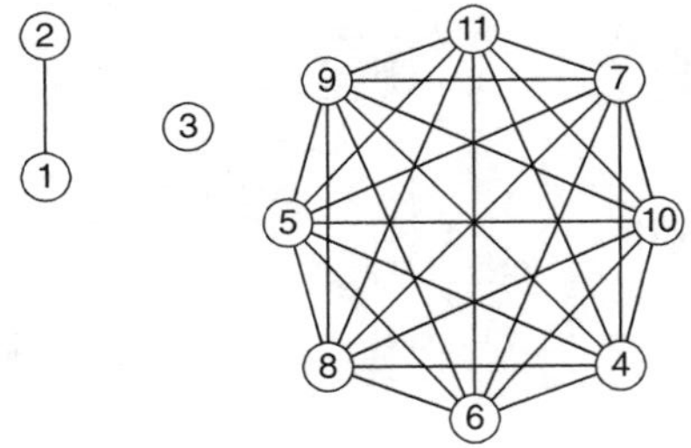

WEB 13: PROTECTED INTERTIDAL

1 – DETRITUS
2 – PLANKTON
3 – ALGAE
4 – ACORN BARNACLES
5 – LIMPETS
6 – OTHER HERBIVOROUS GASTROPODS
7 – M.EDULIS
8 – CHITONS
9 – LITTORINA
10 – THAIS
11 – SEARLESIA
12 – PISASTER
13 – LEPTASTERIAS

FOOD WEB MATRIX

	4	5	6	7	8	9	10	11	12	13
1	1	0	0	1	0	0	0	0	0	0
2	1	0	0	1	0	0	0	0	0	0
3	0	1	1	0	1	1	0	0	0	0
4	0	0	0	0	0	0	1	1	1	1
5	0	0	0	0	0	0	0	1	1	1
6	0	0	0	0	0	0	0	0	1	1
7	0	0	0	0	0	0	1	0	1	1
8	0	0	0	0	0	0	0	1	1	1
9	0	0	0	0	0	0	0	0	1	1
10	0	0	0	0	0	0	0	0	1	1
11	0	0	0	0	0	0	0	0	1	1

CONSUMER OVERLAP $G_R(C;\lambda)$

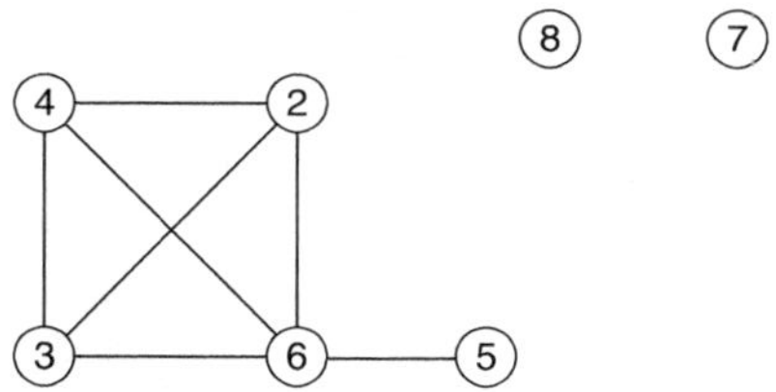

RESOURCE GRAPH $G_C(R;\lambda^{-1})$

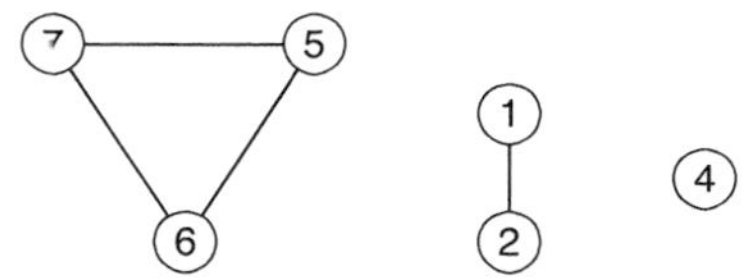

WEB 14: MANGROVE SWAMP

1 – DETRITUS
2 – TENDIPES LARVAE
3 – PROCAMBARUS CLARKII
4 – MOSQUITO LARVAE
5 – XIPHOPHOROUS HELLERI
6 – MELANIA INDEFINITA
7 – LEBISTES RETICULATUS
8 – ELEOSTRIS SANDWICENSIS

FOOD WEB MATRIX

	2	3	4	5	6	7	8
1	1	1	1	0	1	0	0
2	0	0	0	1	1	0	0
4	0	0	0	0	0	1	0
5	0	0	0	0	0	0	1
6	0	0	0	0	0	0	1
7	0	0	0	0	0	0	1

CONSUMER OVERLAP $G_R(C;\lambda)$

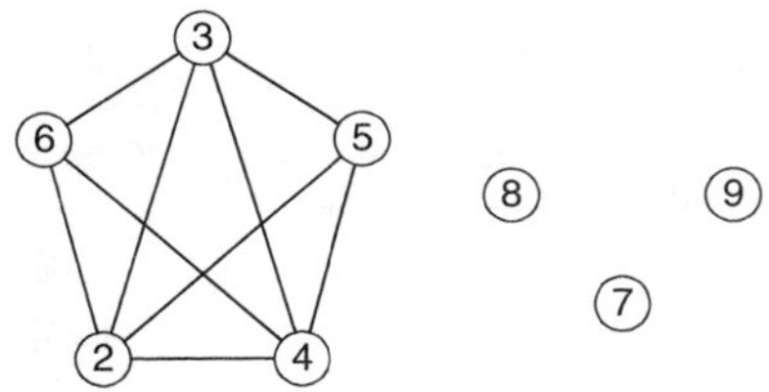

RESOURCE GRAPH $G_C(R;\lambda^{-1})$

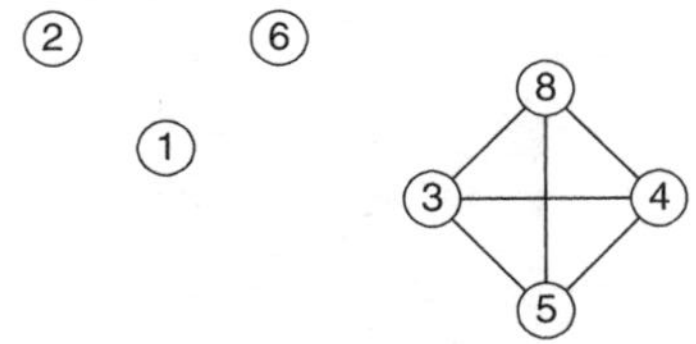

WEB 15: MANGROVE SWAMP

1 – DETRITUS
2 – COPEPODS
3 – XIPHOPHOROUS HELLERI
4 – NERITINA TAHITIENSIS
5 – METOPOGRAPSIS MESSOR
6 – MOSQUITO LARVAE
7 – CHROMOPHOROUS GENIVITTATUS
8 – LEBISTES RETICULATUS
9 – KUHLIA SANDVICENSIS

FOOD WEB MATRIX

	2	3	4	5	6	7	8	9
1	1	1	1	1	1	0	0	0
2	0	0	0	0	0	1	0	0
3	0	0	0	0	0	0	0	1
4	0	0	0	0	0	0	0	1
5	0	0	0	0	0	0	0	1
6	0	0	0	0	0	0	1	0
8	0	0	0	0	0	0	0	1

CONSUMER OVERLAP $G_R(C;\lambda)$

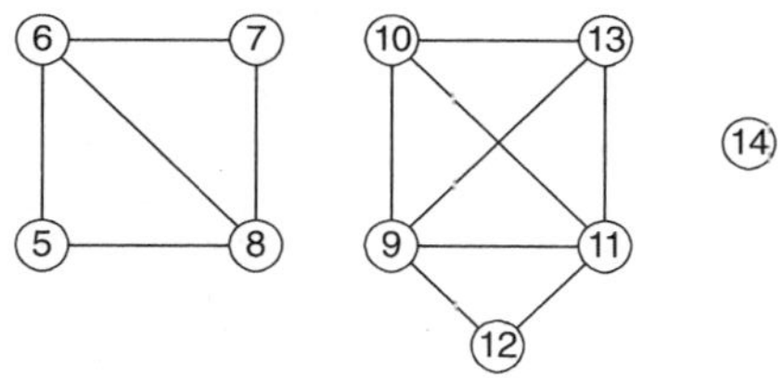

RESOURCE GRAPH $G_C(R;\lambda^{-1})$

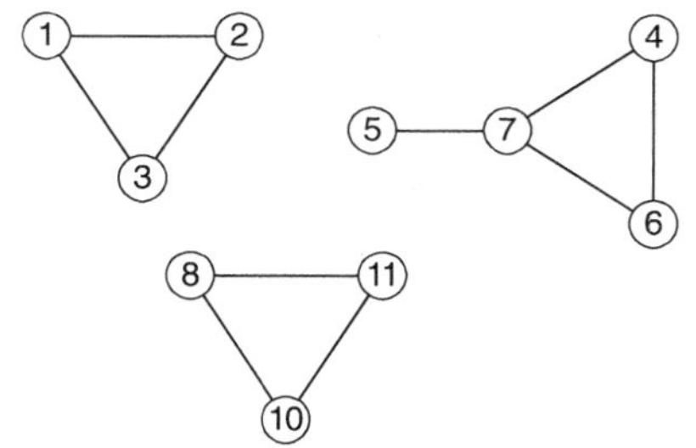

WEB 16: PAMLICO RIVER

1 – DETRITUS
2 – DINOFLAGELLATES
3 – BENTHIC DIATOMS
4 – RUPPIA
5 – ACARTIA TONSA
6 – RANGIA
7 – GAMMARUS AND PALEOMONETES
8 – MUGIL
9 – PARALICHTHYS
10 – BREVOORTIA
11 – LEIOSTOMUS AND MICROPOGON
12 – CALLINECTES
13 – CTENOPHORA
14 – ROCCUS AND CYNOSCION

FOOD WEB MATRIX

	5	6	7	8	9	10	11	12	13	14
1	0	1	0	1	0	0	0	0	0	0
2	1	1	0	1	0	0	0	0	0	0
3	0	1	1	1	0	0	0	0	0	0
4	0	0	0	0	0	0	0	1	0	0
5	0	0	0	0	1	1	1	0	1	0
6	0	0	0	0	0	0	0	1	0	0
7	0	0	0	0	1	0	1	1	0	0
8	0	0	0	0	0	0	0	0	0	1
10	0	0	0	0	0	0	0	0	0	1
11	0	0	0	0	0	0	0	0	0	1

CONSUMER OVERLAP $G_R(C;\lambda)$

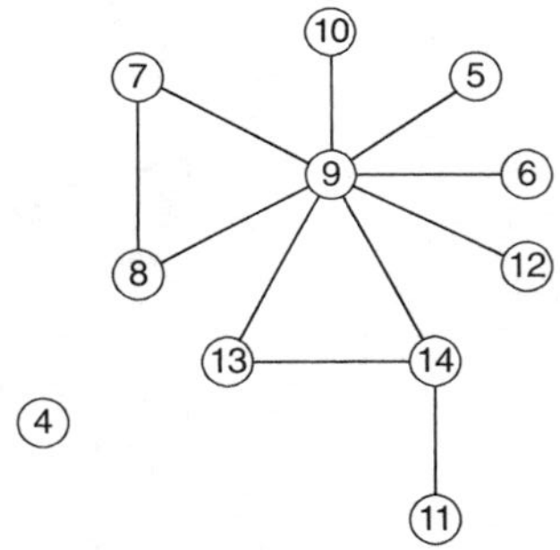

RESOURCE GRAPH $G_C(R;\lambda^{-1})$

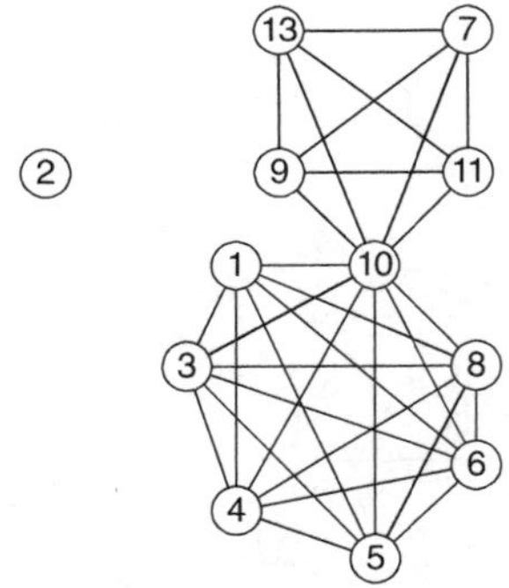

WEB 17: MARSHALLESE REEFS

1 – DETRITUS
2 – PHYTOPLANKTON
3 – BENTHIC ALGAE
4 – ZOOPLANKTON
5 – DETRITUS FEEDERS
6 – ALGAL FEEDERS
7 – SMALL MID–WATER PLANKTON FEEDERS
8 – CORALS
9 – OMNIVORES
10 – SMALL BENTHIC CARNIVORES
11 – LARGE MID–WATER CARNIVORES
12 – CORAL FEEDERS
13 – LARGE BENTHIC CARNIVORES
14 – TRANSIENT CARNIVORES

FOOD WEB MATRIX

	4	5	6	7	8	9	10	11	12	13	14
1	0	1	0	0	0	1	0	0	0	0	0
2	1	0	0	0	0	0	0	0	0	0	0
3	0	0	1	0	0	1	0	0	0	0	0
4	0	0	0	1	1	1	0	0	0	0	0
5	0	0	0	0	0	1	1	0	0	0	0
6	0	0	0	0	0	1	1	0	0	0	0
7	0	0	0	0	0	0	0	1	0	0	1
8	0	0	0	0	0	1	0	0	1	0	0
9	0	0	0	0	0	0	0	0	0	1	1
10	0	0	0	0	0	1	0	0	0	1	1
11	0	0	0	0	0	0	0	0	0	0	1
13	0	0	0	0	0	0	0	0	0	0	1

CONSUMER OVERLAP $G_R(C;\lambda)$

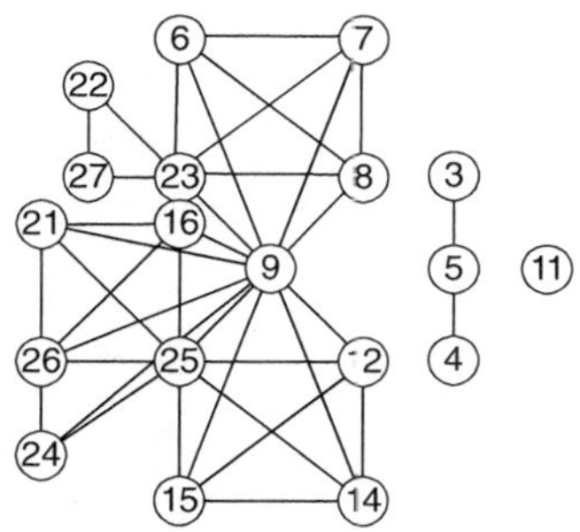

RESOURCE GRAPH $G_C(R;\lambda^{-1})$

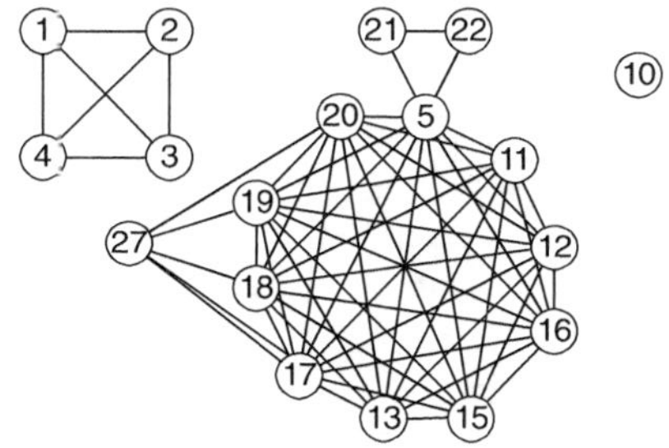

WEB 18: KAPINGAMARANGI ATOLL

1 – ALGAE
2 – PHYTOPLANKTON
3 – ZOOPLANKTON
4 – INVERTEBRATES
5 – FISH
6 – TERNS
7 – FRIGATE BIRDS
8 – BOOBIES
9 – MAN
10 – TURTLE GRASS
11 – SEA TURTLES
12 – PIG
13 – COCONUT
14 – RAT
15 – COCONUT CRABS
16 – FOWL
17 – LAND VEGETATION
18 – CYRTOSPERMA
19 – PANDANUS
20 – BREADFRUIT
21 – INSECTS
22 – SKINKS
23 – REEF HERON
24 – STARLINGS
25 – LANDCRUSTACEA
26 – FUNGI, SNAILS, ANNELIDS
27 – GECKOS

FOOD WEB MATRIX

	3	4	5	6	7	8	9	11	12	14	15	16	21	22	23	24	25	26	27
1	0	0	1	0	0	0	0	0	1	0	0	0	0	0	0	0	0	0	0
2	1	0	1	0	0	0	0	0	0	0	0	0	0	0	0	0	0	0	0
3	0	1	1	0	0	0	0	0	0	0	0	0	0	0	0	0	0	0	0
4	0	0	1	0	0	0	0	0	0	0	0	0	0	0	0	0	0	0	0
5	0	0	0	1	1	1	1	0	0	0	0	0	0	0	1	0	0	0	0
10	0	0	0	0	0	0	0	1	0	0	0	0	0	0	0	0	0	0	0
11	0	0	0	0	0	0	1	0	0	0	0	0	0	0	0	0	0	0	0
12	0	0	0	0	0	0	1	0	0	0	0	0	0	0	0	0	0	0	0
13	0	0	0	0	0	0	1	0	1	1	1	0	0	0	0	0	1	0	0
15	0	0	0	0	0	0	1	0	0	0	0	0	0	0	0	0	0	0	0
16	0	0	0	0	0	0	1	0	0	0	0	0	0	0	0	0	0	0	0
17	0	0	0	0	0	0	1	0	0	0	0	1	1	0	0	0	1	1	0
18	0	0	0	0	0	0	1	0	0	0	0	0	0	0	0	0	1	1	0
19	0	0	0	0	0	0	1	0	0	0	0	0	0	0	0	0	1	1	0
20	0	0	0	0	0	0	1	0	0	0	0	0	0	0	0	0	1	1	0
21	0	0	0	0	0	0	0	0	0	0	0	0	0	0	1	0	0	0	1
22	0	0	0	0	0	0	0	0	0	0	0	0	0	0	1	0	0	0	0
27	0	0	0	0	0	0	0	0	0	0	0	0	0	0	0	0	1	0	0

CONSUMER OVERLAP $G_R(C;\lambda)$

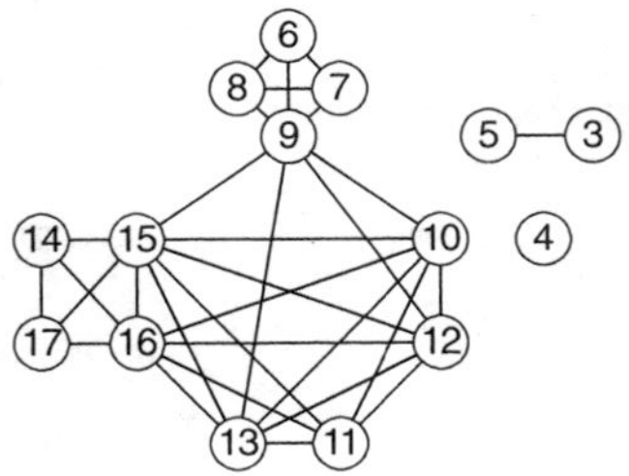

RESOURCE GRAPH $G_C(R;\lambda^{-1})$

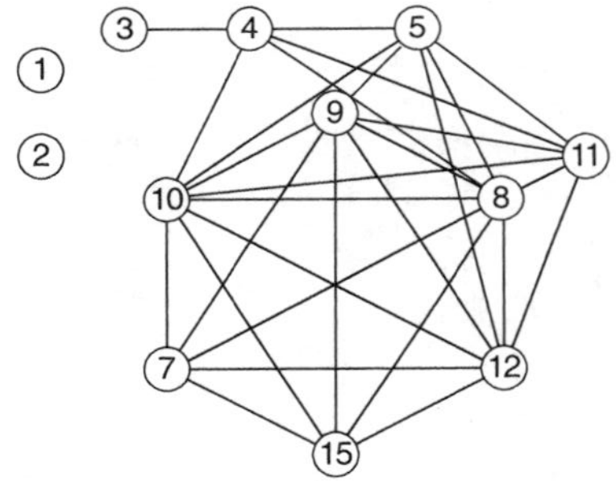

WEB 19: MOOSEHEAD LAKE

1 – PHYTOPLANKTON
2 – BACTERIA
3 – ZOOPLANKTON
4 – CHIRONOMIDS
5 – LITTORAL BROWSERS
6 – CHAOBORUS
7 – LEPTODORA
8 – OSMERUS
9 – CATOSTOMUS
10 – CASTEROSTEUS
11 – CYPRINIDS
12 – COTTUS
13 – PROSOPIUM
14 – SALMO
15 – SALVELINUS
16 – LOTA
17 – CRISTIVOMER

FOOD WEB MATRIX

	3	4	5	6	7	8	9	10	11	12	13	14	15	16	17
1	1	0	1	0	0	0	0	0	0	0	0	0	0	0	0
2	0	1	0	0	0	0	0	0	0	0	0	0	0	0	0
3	0	0	0	1	1	1	1	0	0	0	0	0	0	0	0
4	0	0	0	0	0	0	1	1	0	1	1	0	1	0	0
5	0	0	0	0	0	0	0	1	1	1	1	0	1	1	0
7	0	0	0	0	0	0	0	0	0	0	0	0	0	0	1
8	0	0	0	0	0	0	0	0	0	0	0	1	1	1	1
9	0	0	0	0	0	0	0	0	0	0	0	0	0	1	1
10	0	0	0	0	0	0	0	0	0	0	0	0	1	0	1
11	0	0	0	0	0	0	0	0	0	0	0	0	1	1	0
12	0	0	0	0	0	0	0	0	0	0	0	0	0	1	1
15	0	0	0	0	0	0	0	0	0	0	0	0	0	0	1

CONSUMER OVERLAP $G_R(C;\lambda)$

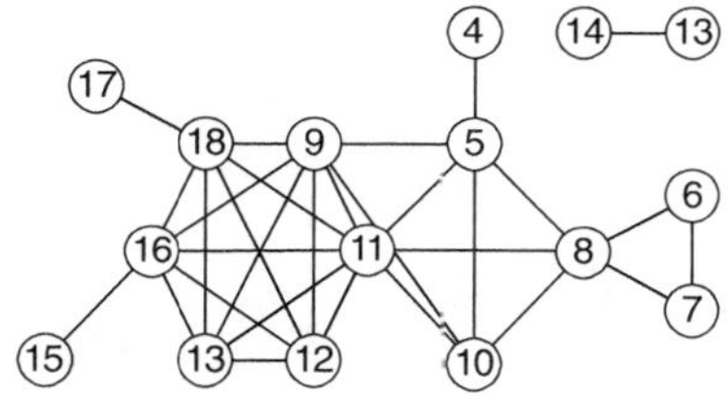

RESOURCE GRAPH $G_C(R;\lambda^{-1})$

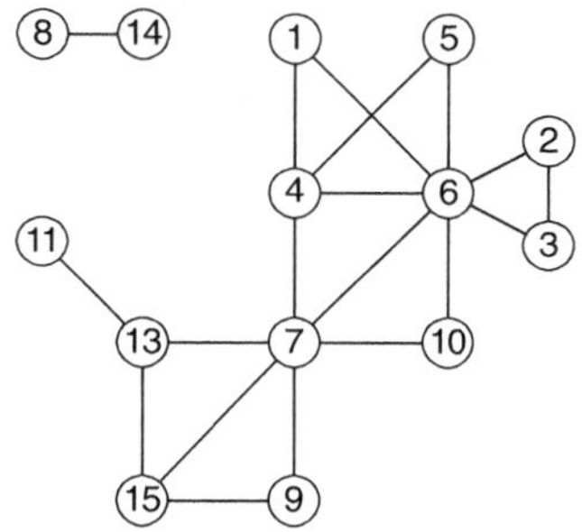

WEB 20: ANTARCTIC PACK ICE

1 – EPONTIC MICROALGAE
2 – PLANKTONIC MICROALGAE
3 – DETRITUS
4 – EPONTIC MICROALGAL FEEDERS
5 – FRY (TREMATOMUS BORCHGREVINKI)
6 – COPEPODS
7 – KRILL (EUPHAUSIA SP.)
8 – BENTHIC FILTER FEEDING INVERTS.
9 – PELAGIC NOTOTHENIDS
10 – PREDATORY PLANKTONIC INVERTS.
11 – CEPHALOPODS
12 – WHALES, CRABEATER SEALS
13 – PENGUINS, PETRELS
14 – BENTHIC CARNIVORES
15 – PREDATORY FISHES
16 – WEDDELL SEALS
17 – TOOTHED WHALES
18 – LEOPARD SEALS
20 – PREDATORY BENTHIC FISHES

FOOD WEB MATRIX

	4	5	6	7	8	9	10	11	12	13	14	15	16	17	18	19
1	1	1	0	0	0	0	0	0	0	0	0	0	0	0	0	0
2	0	0	1	1	1	0	0	0	0	0	0	0	0	0	0	0
3	0	0	0	0	1	0	0	0	0	0	0	0	0	0	0	0
4	0	1	0	0	0	1	1	0	0	0	0	0	0	0	0	0
5	0	0	0	0	0	0	1	0	0	0	0	0	0	0	0	0
6	0	1	0	0	1	0	1	1	0	0	0	0	0	0	0	0
7	0	0	0	0	0	0	0	1	1	1	0	0	1	0	1	0
8	0	0	0	0	0	0	0	0	0	0	1	0	0	0	0	1
9	0	0	0	0	0	0	0	0	0	0	0	1	1	0	0	0
10	0	0	0	0	0	0	0	1	0	0	0	0	0	0	0	0
11	0	0	0	0	0	0	0	0	0	0	0	0	0	1	0	0
13	0	0	0	0	0	0	0	0	0	0	0	0	0	1	1	0
14	0	0	0	0	0	0	0	0	0	0	0	0	0	0	0	1
15	0	0	0	0	0	0	0	0	0	0	0	0	0	0	1	0

CONSUMER OVERLAP $G_R(C;\lambda)$

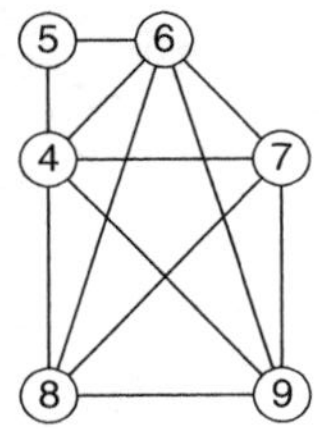

RESOURCE GRAPH $G_C(R;\lambda^{-1})$

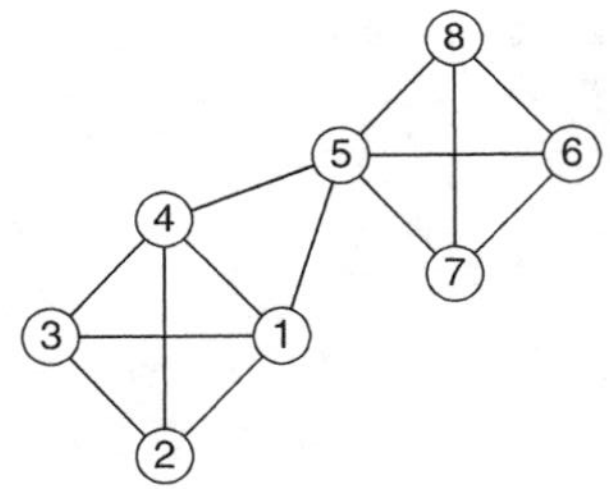

WEB 21: ROSS SEA

1 – ICE ALGAE
2 – PHYTOPLANKTON
3 – DETRITUS
4 – ICE INVERTS.
5 – ZOOPLANKTON
6 – FISH
7 – CEPHALOPODS
8 – PENGUINS
9 – SEALS

FOOD WEB MATRIX

	4	5	6	7	8	9
1	1	1	0	0	0	0
2	0	1	0	0	0	0
3	0	1	0	0	0	0
4	0	1	1	0	0	0
5	1	0	1	1	1	1
6	0	0	0	1	1	1
7	0	0	0	0	1	1
8	0	0	0	0	0	1

CONSUMER OVERLAP $G_R(C;\lambda)$

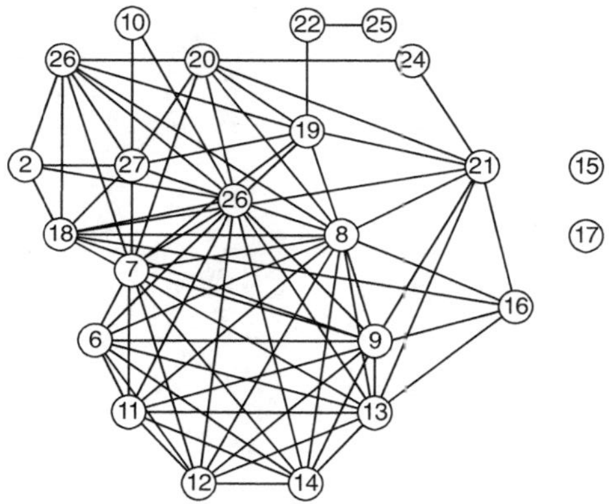

RESOURCE GRAPH $G_C(R;\lambda^{-1})$

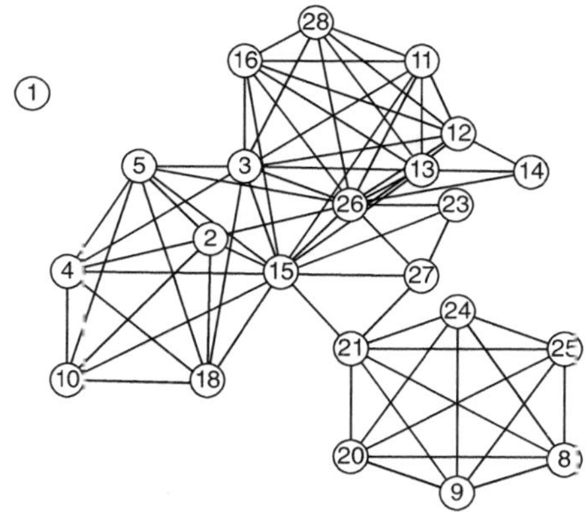

WEB 22: BEAR ISLAND

1 – BACTERIA
2 – PROTOZOA (F.W. PLANKTON)
3 – DEAD PLANTS, PLANTS
4 – ALGAE (F.W. BOTTOM, LITTORAL)
5 – ALGAE (F.W. PLANKTON)
6 – WORMS
7 – GEESE
8 – SNOW BUNTING, PTARMIGAM
9 – PURPLE SANDPIPER
10 – PROTOZOA (F.W. BOTTOM, LITTORAL)
11 – HYMENOPTERA
12 – MITES
13 – DIPTERA (ADULT)
14 – COLLEMBOLA
15 – MOSS
16 – SPIDERS
17 – PROTOZOA (LAND)
18 – ENTOMOSTRACA, ROTIFERA
19 – SKUA, GLAUCOUS GULL
20 – KITTIWAKE GUILLEMOTS, PETREL, AUK
21 – N. EIDER, L.T. DUCK, R.T. DIVER
22 – ARCTIC FOX
23 – MARINE ANIMALS
24 – SEALS
25 – POLAR BEAR
26 – DIPTERA (LARVAE)
27 – ENTOMOSTRACA, ROTIFERA, TARDIGRADA
28 – LEPIDURUS

FOOD WEB MATRIX

	2	6	7	8	9	10	11	12	13	14	15	16	17	18	19	20	21	22	24	25	26	27	28
1	0	0	0	0	0	0	0	0	0	0	0	0	1	0	0	0	0	0	0	0	0	0	0
2	0	0	0	0	0	0	0	0	0	0	0	0	0	1	0	0	0	0	0	0	0	1	1
3	0	1	1	1	1	0	1	1	1	1	0	0	0	0	0	0	0	0	0	0	1	0	0
4	0	0	0	0	0	1	0	0	0	0	0	0	0	0	0	0	0	0	0	0	1	1	0
5	1	0	0	0	0	0	0	0	0	0	0	0	0	1	0	0	0	0	0	0	1	1	1
8	0	0	0	0	0	0	0	0	0	0	0	0	0	0	0	0	0	1	0	0	0	0	0
9	0	0	0	0	0	0	0	0	0	0	0	0	0	0	0	0	0	1	0	0	0	0	0
10	0	0	0	0	0	0	0	0	0	0	0	0	0	0	0	0	0	0	0	0	0	1	0
11	0	0	0	1	1	0	0	0	0	0	0	0	0	0	0	0	0	0	0	0	0	0	0
12	0	0	0	1	1	0	0	0	0	0	0	1	0	0	0	0	0	0	0	0	0	0	0
13	0	0	0	1	1	0	0	0	0	0	0	1	0	0	0	0	0	0	0	0	0	0	0
14	0	0	0	0	0	0	0	0	0	0	0	1	0	0	0	0	0	0	0	0	0	0	0
15	0	0	1	1	0	0	0	0	0	0	0	0	0	0	1	1	0	0	0	0	1	1	1
16	0	0	0	1	1	0	0	0	0	0	0	0	0	0	0	0	0	0	0	0	0	0	0
18	0	0	0	0	0	0	0	0	0	0	0	0	0	0	0	0	0	0	0	0	1	1	1
20	0	0	0	0	0	0	0	0	0	0	0	0	0	0	0	0	0	1	0	0	0	0	0
21	0	0	0	0	0	0	0	0	0	0	0	0	0	0	1	0	0	1	0	0	0	0	0
23	0	0	0	0	0	0	0	0	0	0	0	0	0	0	0	1	1	0	1	0	0	0	0
24	0	0	0	0	0	0	0	0	0	0	0	0	0	0	0	0	0	1	0	1	0	0	0
25	0	0	0	0	0	0	0	0	0	0	0	0	0	0	0	0	0	1	0	0	0	0	0
26	0	0	0	1	1	0	0	0	1	0	0	1	0	1	0	0	1	0	0	0	0	0	0
27	0	0	0	0	0	0	0	0	0	0	0	0	0	0	1	0	1	0	0	0	0	0	0
28	0	0	0	0	1	0	0	0	0	0	0	0	0	0	0	0	0	0	0	0	0	0	0

CONSUMER OVERLAP $G_R(C;\lambda)$

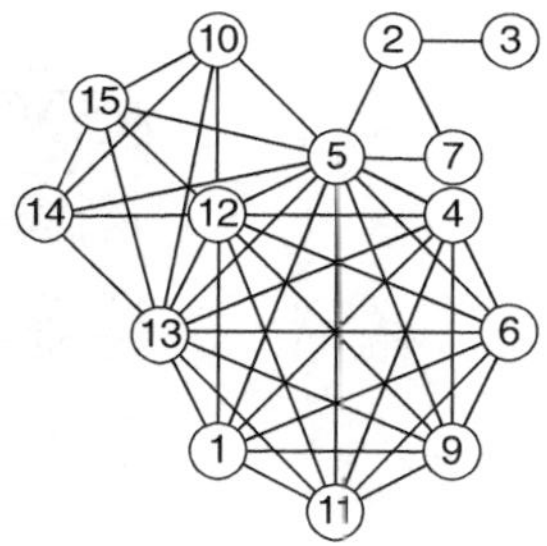

RESOURCE GRAPH $G_C(R;\lambda^{-1})$

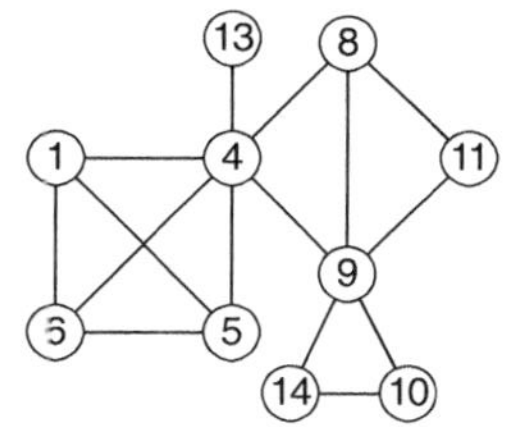

WEB 23: PRAIRIE, CANADA

1 – GROUND SQUIRREL
2 – M. HAWK, COYOTE, R–T. HAWK, WEASEL
3 – BADGER
4 – VOLE
5 – GROUND SQUIRREL (13–STRIPED)
6 – POCKET GOPHER
7 – GREAT HORNED OWL
8 – AGROPYRON, STIPA, HELIANTHUS
9 – SURFACE INSECTS
10 – SPIDERS
11 – SOIL INSECTS
12 – LARKS, SPARROWS, PLOVER
13 – CROW
14 – FROG
15 – GARTER SNAKE

FOOD WEB MATRIX

	1	2	3	4	5	6	7	9	10	11	12	13	14	15
1	0	1	1	0	0	0	0	0	0	0	0	0	0	0
4	0	1	0	0	1	0	1	0	0	0	0	0	0	0
5	0	1	1	0	0	0	0	0	0	0	0	0	0	0
6	0	1	0	0	0	0	0	0	0	0	0	0	0	0
8	1	0	0	1	1	1	0	1	0	1	1	1	0	0
9	0	0	0	0	1	0	0	0	1	0	1	1	1	1
10	0	0	0	0	0	0	0	0	0	0	0	0	1	0
11	0	0	0	0	0	0	0	0	0	0	1	1	0	0
13	0	0	0	0	0	0	1	0	0	0	0	0	0	0
14	0	0	0	0	0	0	0	0	0	0	0	0	1	1

CONSUMER OVERLAP $G_R(C;\lambda)$

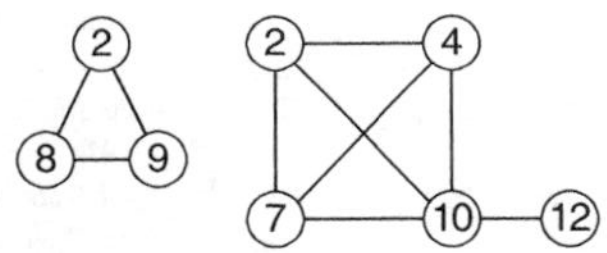

RESOURCE GRAPH $G_C(R;\lambda^{-1})$

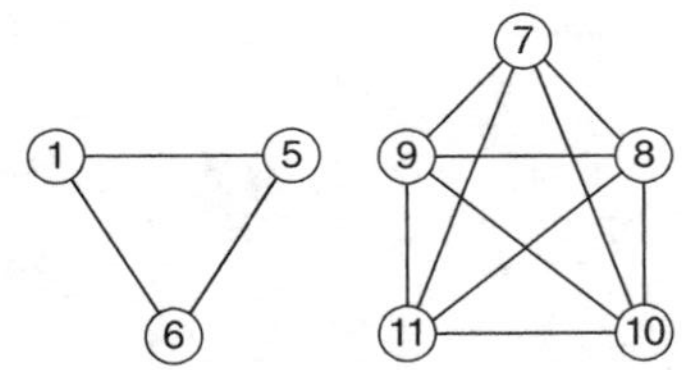

WEB 24: WILLOW FOREST, CANADA

1 – SALIX DISCOLOR
2 – GALERUCELLA DECORA
3 – RW. BLACKBIRD, B. GRACKLE, S. SPARROW
4 – M. YELLOWTHROAT, Y. WARBLER, S. SPARROW
5 – SALIX PETIOLARIS
6 – SALIX LONGIFOLIA
7 – SPIDERS
8 – INSECTS, PONTANIA PETIOLARIDIS
9 – INSECTS, DISYONICHA QUINQUEYITATA
10 – RANA PIPIENS
11 – SNAILS
12 – GARTER SNAKE

FOOD WEB MATRIX

	2	3	4	7	8	9	10	12
1	1	0	0	0	0	1	0	0
5	1	0	0	0	1	1	0	0
6	1	0	0	0	0	1	0	0
7	0	1	1	0	0	0	1	0
8	0	1	1	1	0	0	1	0
9	0	0	1	1	0	0	1	0
10	0	0	0	0	0	0	1	1
11	0	0	0	0	0	0	1	0

CONSUMER OVERLAP $G_R(C;\lambda)$

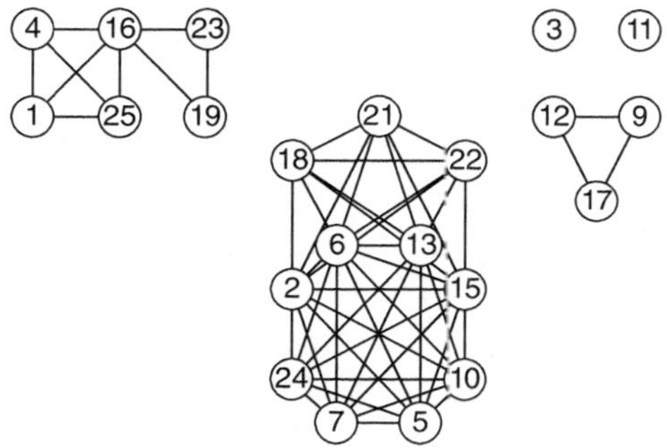

RESOURCE GRAPH $G_C(R;\lambda^{-1})$

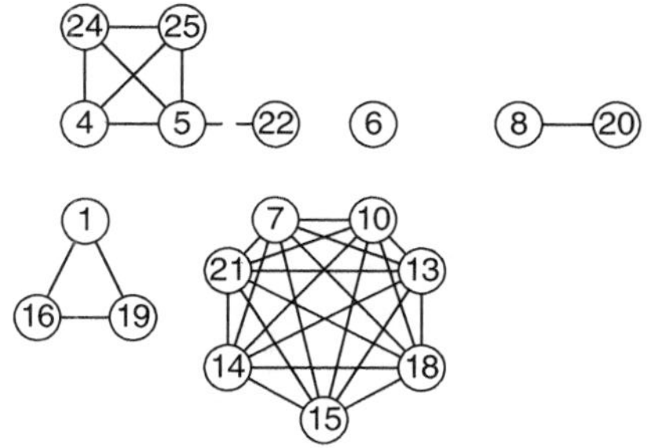

WEB 25: ASPEN FOREST

1 – B. ORIOLE, CHICKADEE, L. FLYCATCHR
2 – CANKER, FOMES
3 – HAIRY, DOWNY WOODPECKERS
4 – SPIDERS (MATURE FOREST)
5 – INSECTS (MATURE FOREST)
6 – DICERA, SAPERDA
7 – RED SQUIRREL
8 – POPULUS, CORNUS, CORYLUS, PYROLA
9 – GOSHAWK
10 – REDBACKED VOLE
11 – COOPERS, SHARPSHINNED HAWKS
12 – GREAT HORNED OWL
13 – RUFFED GROUSE
14 – FLICKER
15 – CROW
16 – HOUSE WREN
17 – TICKS
18 – SNOWSHOE RABBIT
19 – R.E. VIREO, Y. WARBLER, G. FINCH
20 – POPULUS, SYMPHORICARPOS, PRUNUS
21 – FRANKLIN GROUND SQUIRREL
22 – INSECTS (FOREST EDGE)
23 – SPIDERS (FOREST EDGE)
24 – SNAILS
25 – FROGS

FOOD WEB MATRIX

	1	2	3	4	5	6	7	9	10	11	12	13	15	16	17	18	19	21	22	23	24	25
1	0	0	0	0	0	0	0	0	0	1	0	0	0	0	0	0	0	0	0	0	0	0
4	1	0	0	0	0	0	0	0	0	0	0	0	0	0	0	0	0	0	0	0	0	1
5	1	0	0	1	0	0	0	0	0	0	0	0	0	1	0	0	0	0	0	0	0	1
6	0	0	1	0	0	0	0	0	0	0	0	0	0	0	0	0	0	0	0	0	0	0
7	0	0	0	0	0	0	0	0	0	0	1	0	0	0	0	0	0	0	0	0	0	0
8	0	1	0	0	1	1	1	0	1	0	0	1	1	0	0	0	0	0	0	0	1	0
10	0	0	0	0	0	0	0	0	0	0	1	0	0	0	0	0	0	0	0	0	0	0
13	0	0	0	0	0	0	0	1	0	0	1	0	0	0	0	0	0	0	0	0	0	0
14	0	0	0	0	0	0	0	0	0	0	1	0	0	0	0	0	0	0	0	0	0	0
15	0	0	0	0	0	0	0	0	0	0	1	0	0	0	0	0	0	0	0	0	0	0
16	0	0	0	0	0	0	0	0	0	1	0	0	0	0	0	0	0	0	0	0	0	0
18	0	0	0	0	0	0	0	1	0	0	1	0	0	0	1	0	0	0	0	0	0	0
19	0	0	0	0	0	0	0	0	0	1	0	0	0	0	0	0	0	0	0	0	0	0
20	0	1	0	0	0	1	0	0	0	0	0	1	1	0	0	1	0	1	1	0	0	0
21	0	0	0	0	0	0	0	0	0	0	1	0	0	0	0	0	0	0	0	0	0	0
22	0	0	0	0	0	0	0	0	0	0	0	0	0	1	0	0	1	0	0	1	0	0
24	0	0	0	0	0	0	0	0	0	0	0	0	0	0	0	0	0	0	0	0	0	1
25	0	0	0	0	0	0	0	0	0	0	0	0	0	0	0	0	0	0	0	0	0	1

CONSUMER OVERLAP $G_R(C;\lambda)$

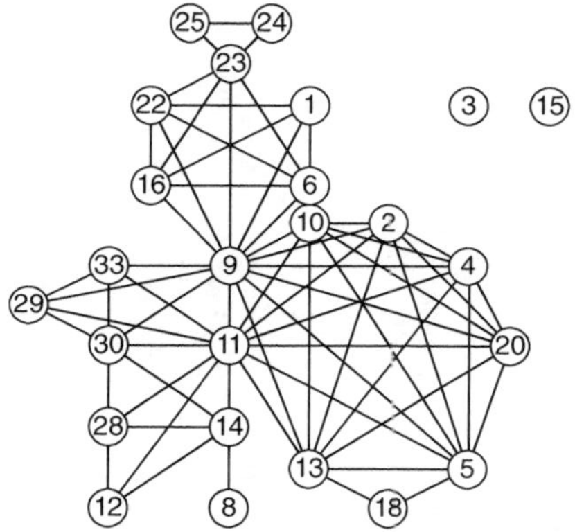

RESOURCE GRAPH $G_C(R;\lambda^{-1})$

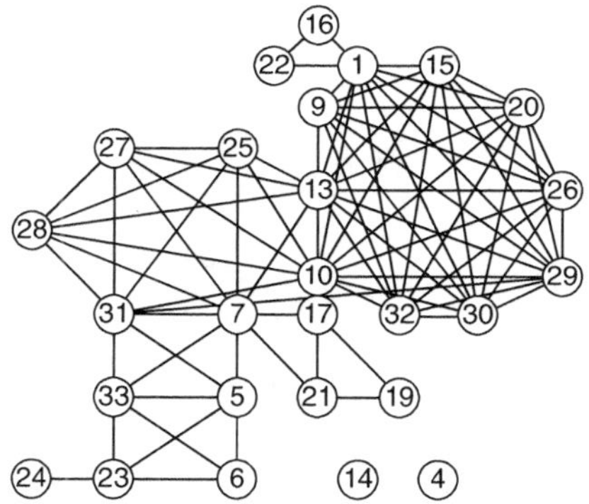

WEB 26: ASPEN PARKLAND

1 – B. ORIOLE, CHICKADEE, L.F–CATCHER, R.G–BEAK, W. THRUSH
2 – CANKER, FOMES
3 – HAIRY, DOWNY WOODPECKERS
4 – DICERCA, SAPERDA
5 – INSECTS
6 – SPIDERS
7 – POPULUS, CORNUS, CORYLUS, PYROLA, ARALIA
8 – COOPERS, SHARPSHINNED HAWKS
9 – CROW
10 – RUFFED GROUSE
11 – MAN
12 – GOSHAWK
13 – RABBIT
14 – GREAT HORNED OWL
15 – FLICKER
16 – M. YELLOWTHROAT, Y. WARBLER, S. SPARROW
17 – SALIX LONGIFOLIA
18 – GALERUCELLA DECORA
19 – SALIX DISCOLOR
20 – RED SQUIRREL
21 – SALIX PETIOLARIS
22 – Y. WARBLER, R. BLACKBIRD, B. GRACKLE
23 – RANA PIPIENS
24 – GARTER SNAKE
25 – FISH
26 – COOTS
27 – DUCKS
28 – COYOTE, WEASEL, SKUNK
29 – PRAIRIE VOLE
30 – POCKET GOPHERS, GROUND SQUIRRELS
31 – AGROSTIS, AGROPYRON, STIPA, HELIANTHUS
32 – MICE
33 – CUTWORMS, GRASSHOPPERS, CLICKBEETLES
34 – ANTS

FOOD WEB MATRIX

	1	2	3	4	5	6	8	9	10	11	12	13	14	15	16	18	20	22	23	24	25	26	29	30	33
1	0	0	0	0	0	0	1	0	0	0	0	0	1	0	0	0	0	0	0	0	0	0	0	0	0
4	0	0	1	0	0	0	0	0	0	0	0	0	0	0	0	0	0	0	0	0	0	0	0	0	0
5	1	0	0	0	0	1	0	1	0	0	0	0	0	0	1	0	0	1	1	0	0	0	0	0	0
6	1	0	0	0	0	0	0	0	0	0	0	0	0	0	1	0	0	0	1	0	0	0	0	0	0
7	0	1	0	1	1	0	0	1	1	1	0	1	0	0	0	0	1	0	0	0	0	0	0	0	0
9	0	0	0	0	0	0	0	0	0	0	0	0	1	0	0	0	0	0	0	0	0	0	0	0	0
10	0	0	0	0	0	0	0	0	0	1	1	0	1	0	0	0	0	0	0	0	0	0	0	0	0
13	0	0	0	0	0	0	0	0	0	1	1	0	1	0	0	0	0	0	0	0	0	1	0	0	0
14	0	0	0	0	0	0	0	0	0	0	0	0	0	0	0	0	0	0	0	0	0	0	0	0	0
15	0	0	0	0	0	0	0	0	0	0	0	0	1	0	0	0	0	0	0	0	0	0	0	0	0
16	0	0	0	0	0	0	1	0	0	0	0	0	0	0	0	0	0	0	0	0	0	0	0	0	0
17	0	0	0	0	1	0	0	0	0	0	0	1	0	0	0	1	0	0	0	0	0	0	0	0	0
19	0	0	0	0	0	0	0	0	0	0	0	0	0	0	0	1	0	0	0	0	0	0	0	0	0
20	0	0	0	0	0	0	0	0	0	0	0	0	1	0	0	0	0	0	0	0	0	0	0	0	0
21	0	0	0	0	1	0	0	0	0	0	0	1	0	0	0	1	0	0	0	0	0	0	0	0	0
22	0	0	0	0	0	0	1	0	0	0	0	0	0	0	0	0	0	0	0	0	0	0	0	0	0
23	0	0	0	0	0	0	0	0	0	0	0	0	0	0	0	0	0	0	1	1	1	0	0	0	0
24	0	0	0	0	0	0	0	0	0	0	0	0	0	0	0	0	0	0	0	0	1	0	0	0	0
25	0	0	0	0	0	0	0	0	0	1	0	0	0	0	0	0	0	0	0	0	0	0	0	0	0
26	0	0	0	0	0	0	0	0	0	0	0	0	1	0	0	0	0	0	0	0	0	0	0	0	0
27	0	0	0	0	0	0	0	0	0	1	0	0	0	0	0	0	0	0	0	0	0	0	0	0	0
28	0	0	0	0	0	0	0	0	0	1	0	0	0	0	0	0	0	0	0	0	0	0	0	0	0
29	0	0	0	0	0	0	0	0	0	0	0	0	1	0	0	0	0	0	0	0	0	1	0	1	0
30	0	0	0	0	0	0	0	0	0	0	0	0	1	0	0	0	0	0	0	0	0	1	0	0	0
31	0	0	0	0	0	0	0	1	0	1	0	0	0	0	0	0	0	0	0	0	0	0	1	1	1
32	0	0	0	0	0	0	0	0	0	0	0	0	1	0	0	0	0	0	0	0	0	0	0	0	0
33	0	0	0	0	0	0	0	1	0	0	0	0	0	0	0	0	0	0	1	0	0	0	0	0	0
34	0	0	0	0	0	0	0	0	0	0	0	0	0	1	0	0	0	0	0	0	0	0	0	0	0

CONSUMER OVERLAP $G_R(C;\lambda)$

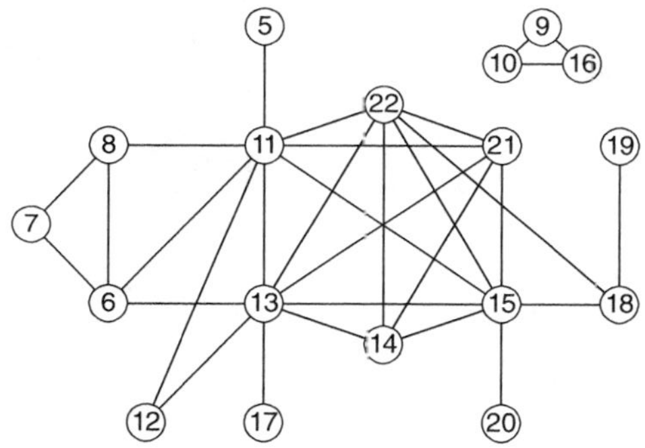

RESOURCE GRAPH $G_C(R;\lambda^{-1})$

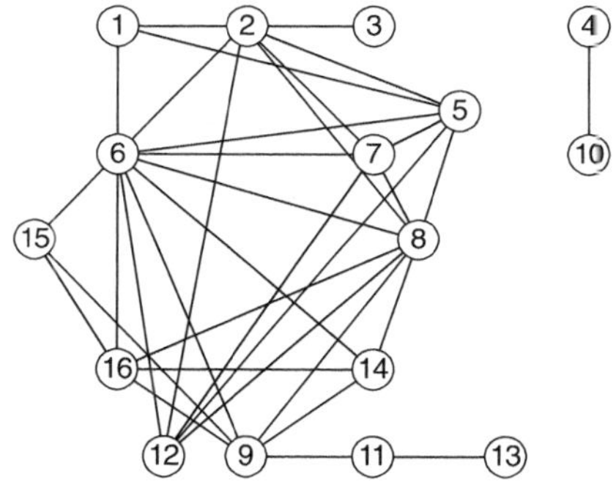

WEB 27: WHYTAM WOOD

1 – WEASELS
2 – TREES AND BUSHES
3 – OAK TREES
4 – TOTAL LITTER
5 – INSECTS
6 – WINTER MOTH
7 – TARTRIX
8 – OTHER LEAF FEEDERS
9 – EARTHWORMS
10 – FUNGI
11 – VOLES, NICE
12 – SPIDERS
13 – TITMICE
14 – CYZENIS
15 – PHILANTHUS, ABAX, ETC.
16 – SOIL INSECTS, MITES
17 – PARASITES
18 – OWLS
19 – WEASELS
20 – HYPERPARASITES
21 – SHREWS
22 – MOLES

FOOD WEB MATRIX

	5	6	7	8	9	10	11	12	13	14	15	16	17	18	19	20	21	22
1	1	0	0	0	0	0	1	0	0	0	0	0	0	0	0	0	0	0
2	0	1	0	1	0	0	1	0	1	0	0	0	0	0	0	0	0	0
3	0	1	1	1	0	0	0	0	0	0	0	0	0	0	0	0	0	0
4	0	0	0	0	1	1	0	0	0	0	0	1	0	0	0	0	0	0
5	0	0	0	0	0	0	1	1	1	0	0	0	0	0	0	0	0	0
6	0	0	0	0	0	0	1	0	1	1	1	0	0	0	0	0	1	1
7	0	0	0	0	0	0	0	0	1	0	0	0	0	0	0	0	0	0
8	0	0	0	0	0	0	0	0	1	0	1	0	0	0	0	0	0	0
9	0	0	0	0	0	0	0	0	0	0	1	0	0	1	0	0	0	1
10	0	0	0	0	0	0	0	0	0	0	0	1	0	0	0	0	0	0
11	0	0	0	0	0	0	0	0	0	0	0	0	0	1	1	0	0	0
12	0	0	0	0	0	0	0	0	1	0	0	0	1	0	0	0	0	0
13	0	0	0	0	0	0	0	0	0	0	0	0	0	0	1	0	0	0
14	0	0	0	0	0	0	0	0	0	0	1	0	0	0	0	1	0	0
15	0	0	0	0	0	0	0	0	0	0	0	0	0	0	0	0	1	1
16	0	0	0	0	0	0	0	0	0	0	1	0	0	0	0	0	1	0

CONSUMER OVERLAP $G_R(C;\lambda)$

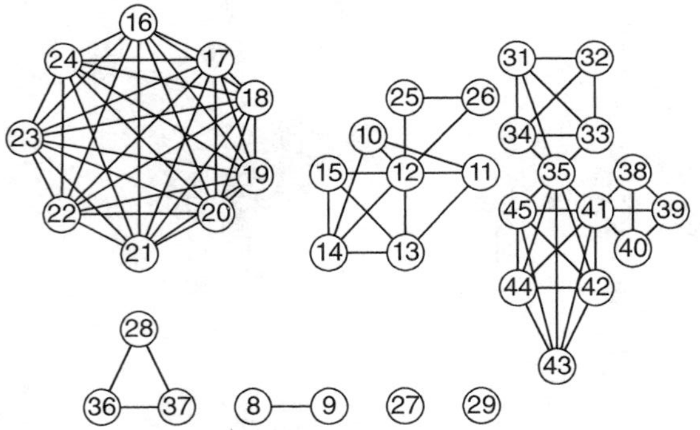

RESOURCE GRAPH $G_C(R;\lambda^{-1})$

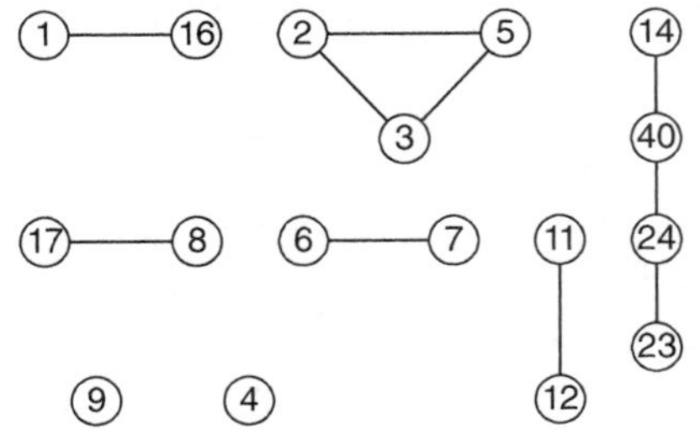

WEB 28: N.Z. SALT MEADOW

1 – HUMUS
2 – LEAVES
3 – FLOWERS
4 – SEEDS
5 – ROOTS
6 – ALGAE
7 – BACTERIA
8 – PROTOZOA, ROTIFERS
9 – NEMATODES
10 – WEEVIL LARVAE
11 – COCCIDS
12 – LEPIDOPTERAN LARVAE
13 – OTHER HEMIPTERA
14 – RABBITS
15 – RUST
16 – FUNGI
17 – COLLEMBOLA
18 – HARPACTICOIDS
19 – STAPHYLINIDS
20 – DIPTERAN LARVAE
21 – HAPLOTAXID WORMS
22 – ORIBATIDS
23 – OTHER MITES
24 – AMPHIPODS
25 – BUMBLEBEES
26 – OTHER HYMENOPTERANS
27 – REDPOLLS
28 – CARNIVOROUS NEMATODES
29 – TARDIGRADES
30 – PARASITIC HYMEN, LARVAE
31 – T. RETORTAEFORMIS
32 – G. STRIGOSUM
33 – P. AMBIGUUS
34 – STOATS
35 – HARRIER HAWKS
36 – TROMBIDIFORM MITES
37 – SPIDERS
38 – ANTS
39 – STARLINGS
40 – DOTTEREL
41 – HYMEN. CESTODES
42 – NEMATODE ECHINURIA
43 – ANALGESID MITES
44 – OTHER MITES
45 – LICE

FOOD WEB MATRIX

	8	9	10	11	12	13	14	15	16	17	18	19	20	21	22	23	24	25	26	27	28	29	30	31	32	33	34	35	36	37	38	39	40	41	42	43	44	45
1	0	0	0	0	0	0	0	0	1	1	1	1	1	1	1	1	1	0	0	0	0	0	0	0	0	0	0	0	0	0	0	0	0	0	0	0	0	0
2	0	0	1	0	1	1	1	1	0	0	0	0	0	0	0	0	0	0	0	0	0	0	0	0	0	0	0	0	0	0	0	0	0	0	0	0	0	0
3	0	0	0	0	1	0	0	0	0	0	0	0	0	0	0	0	0	1	1	0	0	0	0	0	0	0	0	0	0	0	0	0	0	0	0	0	0	0
4	0	0	0	0	0	0	0	0	0	0	0	0	0	0	0	0	0	0	0	1	0	0	0	0	0	0	0	0	0	0	0	0	0	0	0	0	0	0
5	0	0	1	1	1	1	0	0	0	0	0	0	0	0	0	0	0	0	0	0	0	0	0	0	0	0	0	0	0	0	0	0	0	0	0	0	0	0
6	0	1	0	0	0	0	0	0	0	0	0	0	0	0	0	0	0	0	0	0	0	0	0	0	0	0	0	0	0	0	0	0	0	0	0	0	0	0
7	1	1	0	0	0	0	0	0	0	0	0	0	0	0	0	0	0	0	0	0	0	0	0	0	0	0	0	0	0	0	0	0	0	0	0	0	0	0
8	0	0	0	0	0	0	0	0	0	0	0	0	0	0	0	0	0	0	0	0	1	0	0	0	0	0	0	0	0	0	0	0	0	0	0	0	0	0
9	0	0	0	0	0	0	0	0	0	0	0	0	0	0	0	0	0	0	0	0	0	1	0	0	0	0	0	0	0	0	0	0	0	0	0	0	0	0
11	0	0	0	0	0	0	0	0	0	0	0	0	0	0	0	0	0	0	0	0	0	0	1	0	0	0	0	0	0	0	0	0	0	0	0	0	0	0
12	0	0	0	0	0	0	0	0	0	0	0	0	0	0	0	0	0	0	0	0	0	0	1	0	0	0	0	0	0	0	0	0	0	0	0	0	0	0
14	0	0	0	0	0	0	0	0	0	0	0	0	0	0	0	0	0	0	0	0	0	0	0	1	1	1	1	1	0	0	0	0	0	0	0	0	0	0
16	0	0	0	0	0	0	0	0	0	1	0	0	0	0	0	0	0	0	0	0	0	0	0	0	0	0	0	0	0	0	0	0	0	0	0	0	0	0
17	0	0	0	0	0	0	0	0	0	0	0	0	0	0	0	0	0	0	0	0	1	0	0	0	0	0	0	0	1	1	0	0	0	0	0	0	0	0
23	0	0	0	0	0	0	0	0	0	0	0	0	0	0	0	0	0	0	0	0	0	0	0	0	0	0	0	0	0	0	1	0	0	0	0	0	0	0
24	0	0	0	0	0	0	0	0	0	0	0	0	0	0	0	0	0	0	0	0	0	0	0	0	0	0	0	0	0	0	1	1	1	1	0	0	0	0
40	0	0	0	0	0	0	0	0	0	0	0	0	0	0	0	0	0	0	0	0	0	0	0	0	0	0	0	1	0	0	0	0	0	1	1	1	1	1

CONSUMER OVERLAP $G_R(C;\lambda)$

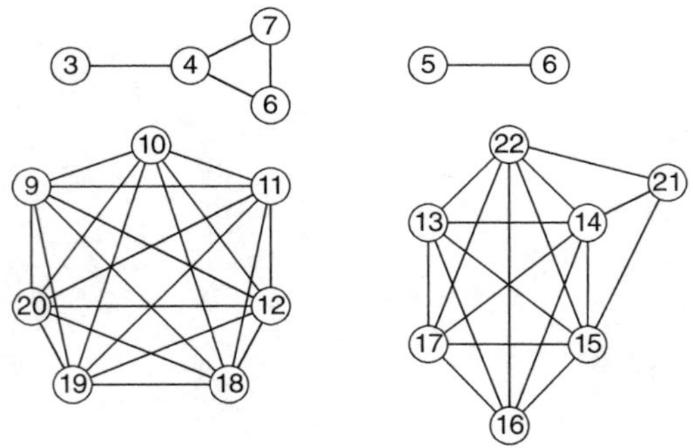

RESOURCE GRAPH $G_C(R;\lambda^{-1})$

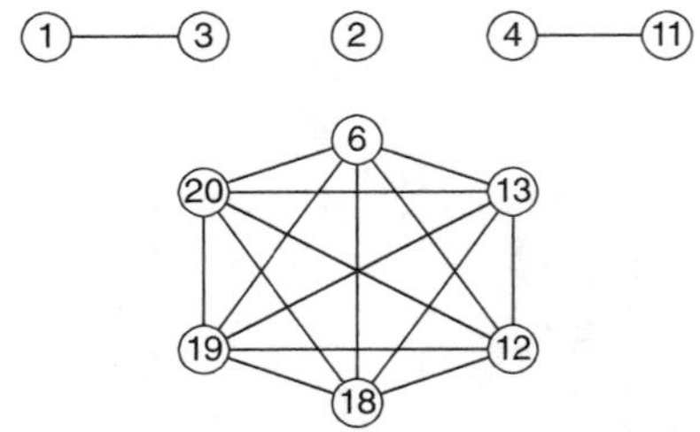

WEB 29: ARCTIC SEAS

1 – PHYTOPLANKTON
2 – DETRITUS
3 – ZOOPLANKTON
4 – ZOOPLANKTON
5 – BACTERIA
6 – BENTHIC INVERTS.
7 – RIGHT WHALES
8 – CLUPEID FISHES
9 – RORQUALS
10 – ARCTIC CHAR
11 – CAPLIN
12 – COD
13 – BENTHIC VERTEBRATES
14 – BEARDED SEAL
15 – BELUGA
16 – NARWHAL
17 – WALRUS
18 – HARP SEAL
19 – HARBOUR SEAL
20 – RINGED SEAL
21 – KILLER WHALE
22 – GREENLAND SHARK

FOOD WEB MATRIX

	3	4	5	6	7	8	9	10	11	12	13	14	15	16	17	18	19	20	21	22
1	1	1	0	0	0	0	0	0	0	0	0	0	0	0	0	0	0	0	0	0
2	0	0	1	1	0	0	0	0	0	0	0	0	0	0	0	0	0	0	0	0
3	0	1	0	0	1	1	0	0	0	0	0	0	0	0	0	0	0	0	0	0
4	0	0	0	0	0	0	1	1	1	1	0	0	0	0	0	1	1	1	0	0
6	0	0	0	0	0	0	0	0	0	0	1	1	1	1	1	0	0	0	0	1
11	0	0	0	0	0	0	0	0	0	1	0	0	0	0	0	1	1	0	0	0
12	0	0	0	0	0	0	0	0	0	0	0	0	0	0	0	0	0	0	1	1
13	0	0	0	0	0	0	0	0	0	0	0	1	1	0	0	0	0	0	1	1
18	0	0	0	0	0	0	0	0	0	0	0	0	0	0	0	0	0	0	1	1
19	0	0	0	0	0	0	0	0	0	0	0	0	0	0	0	0	0	0	1	1
20	0	0	0	0	0	0	0	0	0	0	0	0	0	0	0	0	0	0	1	1

CONSUMER OVERLAP $G_R(C;\lambda)$

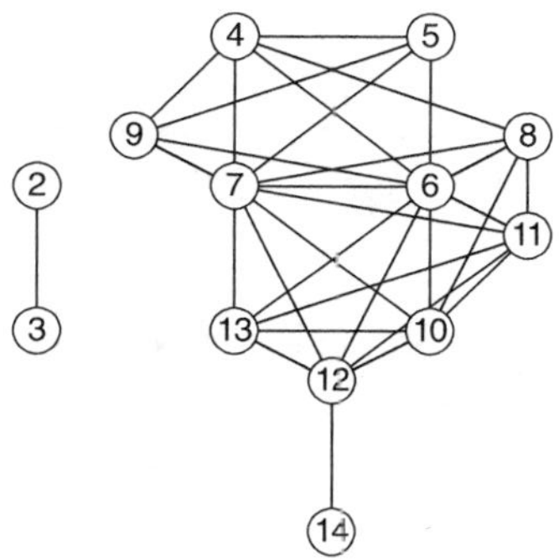

RESOURCE GRAPH $G_C(R;\lambda^{-1})$

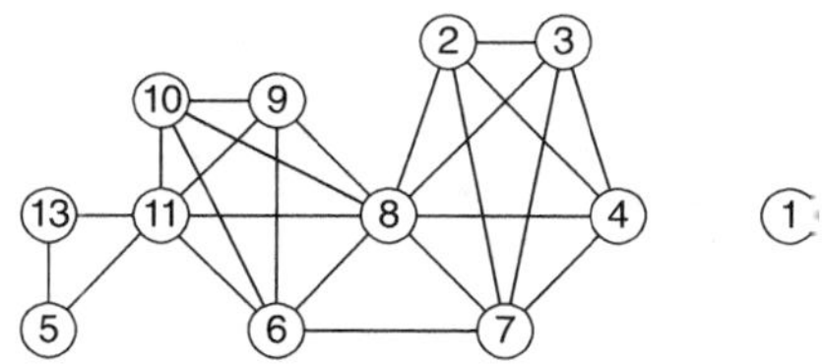

WEB 30: ANTARCTIC SEAS

1 – DETRITUS
2 – SMALL PHYTOPLANKTON+ SAPRO.
3 – MEDIUM PHYTOPLANKTON+ SAPRO.
4 – LARGE PHYTOPLANKTON+ SAPRO.
5 – OIKOPLEURA
6 – COPEPODITES
7 – PARACALANUS
8 – INVERT. LARVAE
9 – NAUPLIUSES
10 – CALANUS (COPE.)
11 – OMNIVORES
12 – PRIMARY CARNIVORES
13 – SECONDARY CARNIVORES
14 – TERTIARY CARNIVORES

FOOD WEB MATRIX

	2	3	4	5	6	7	8	9	10	11	12	13	14
1	1	1	0	0	0	0	0	0	0	0	0	0	0
2	0	0	1	1	1	1	0	1	0	0	0	0	0
3	0	0	1	0	1	1	1	0	0	0	0	0	0
4	0	0	0	0	1	1	1	0	0	0	0	0	0
5	0	0	0	0	0	0	0	0	0	0	0	0	1
6	0	0	0	0	0	0	0	0	1	0	1	0	0
7	0	0	0	0	1	0	1	0	1	1	0	0	0
8	0	0	0	0	1	1	0	0	1	1	1	1	0
9	0	0	0	0	0	0	0	0	0	0	1	0	0
10	0	0	0	0	0	0	0	0	0	0	1	0	0
11	0	0	0	0	0	0	0	0	0	0	1	0	1
13	0	0	0	0	0	0	0	0	0	0	0	0	1

CONSUMER OVERLAP $G_R(C;\lambda)$

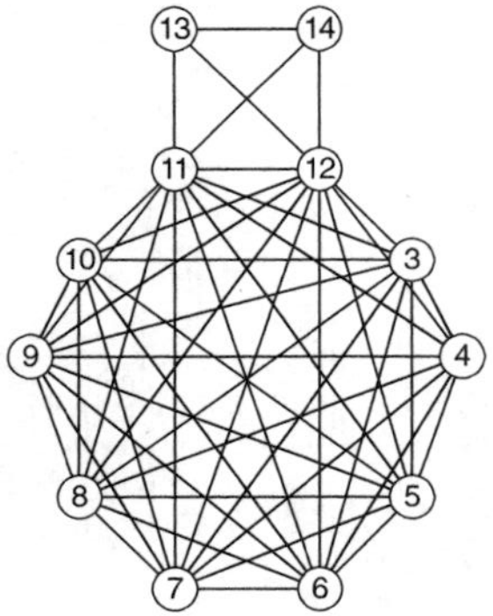

RESOURCE GRAPH $G_C(R;\lambda^{-1})$

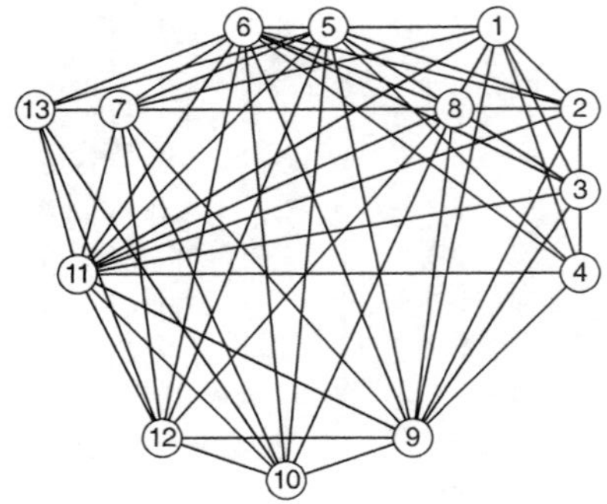

WEB 31: BLACK SEA, EPIPLANKTON

1 – DETRITUS
2 – SMALL PHYTOPLANKTON
3 – MEDIUM PHYTOPLANKTON
4 – LARGE PRODUCERS
5 – OIKOPLEURA
6 – COPEPODITES
7 – PARACALANUS
8 – INVERT. LARVAE
9 – NAUPLIUSES
10 – PSEUDOCALANUS
11 – OMNIVORES
12 – PRIMARY CARNIVORES
13 – SECONDARY CARNIVORES
14 – TERTIARY CARNIVORES

FOOD WEB MATRIX

	3	4	5	6	7	8	9	10	11	12	13	14
1	1	1	1	1	1	1	1	1	1	1	0	0
2	0	0	1	1	1	1	1	1	1	0	0	0
3	0	0	1	1	0	1	1	1	1	0	0	0
4	0	0	0	1	0	1	1	1	1	0	0	0
5	0	0	0	0	0	0	0	0	1	1	1	1
6	0	0	0	0	0	0	0	0	1	1	1	1
7	0	0	0	0	0	0	0	0	0	1	1	1
8	0	0	0	0	0	0	0	0	1	0	1	0
9	0	0	0	0	0	0	0	0	1	1	1	0
10	0	0	0	0	0	0	0	0	0	0	1	1
11	0	0	0	0	0	0	0	0	1	0	1	1
12	0	0	0	0	0	0	0	0	0	0	1	1
13	0	0	0	0	0	0	0	0	0	0	0	1

CONSUMER OVERLAP $G_R(C;\lambda)$

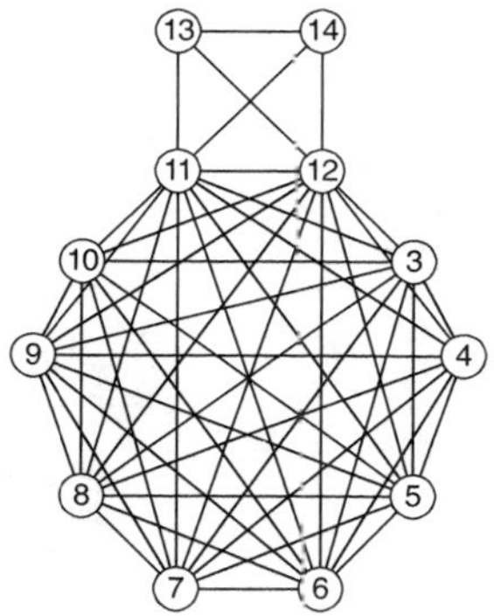

RESOURCE GRAPH $G_C(R;\lambda^{-1})$

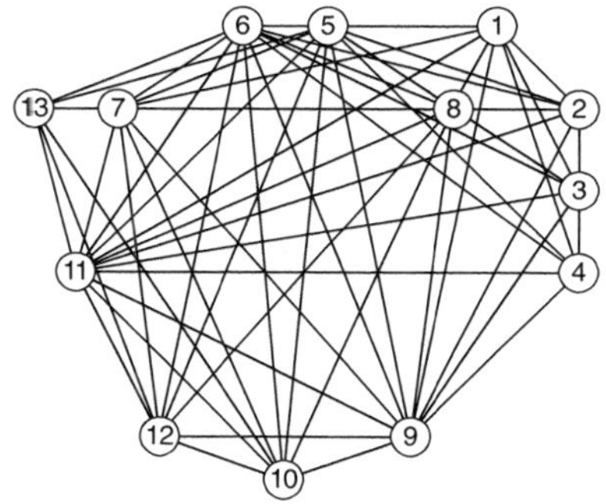

WEB 32: BLACK SEA, BATH PLANK

1 – DETRITUS
2 – SMALL PHYTOPLANKTON
3 – MEDIUM PHYTOPLANKTON
4 – LARGE PRODUCERS
5 – OIKOPLEURA
6 – COPEPODITES
7 – PARACALANUS
8 – INVERT. LARVAE
9 – NAUPLIUSES
10 – PSEUDOCALANUS
11 – OMNIVORES
12 – PRIMARY CARNIVORES
13 – SECONDARY CARNIVORES
14 – TERTIARY CARNIVORES

FOOD WEB MATRIX

	3	4	5	6	7	8	9	10	11	12	13	14
1	1	1	1	1	1	1	1	1	1	1	0	0
2	0	0	1	1	1	1	1	1	1	0	0	0
3	0	0	1	1	0	1	1	1	1	0	0	0
4	0	0	0	1	0	1	1	1	1	0	0	0
5	0	0	0	0	0	0	0	0	1	1	1	1
6	0	0	0	0	0	0	0	0	1	1	1	1
7	0	0	0	0	0	0	0	0	0	1	1	1
8	0	0	0	0	0	0	0	0	1	0	1	0
9	0	0	0	0	0	0	0	0	1	1	1	0
10	0	0	0	0	0	0	0	0	0	0	1	1
11	0	0	0	0	0	0	0	0	1	0	1	1
12	0	0	0	0	0	0	0	0	0	0	1	1
13	0	0	0	0	0	0	0	0	0	0	0	1

CONSUMER OVERLAP $G_R(C;\lambda)$

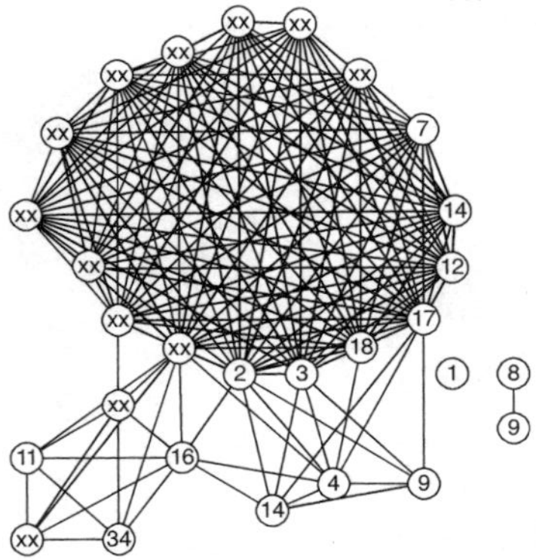

RESOURCE GRAPH $G_C(R;\lambda^{-1})$

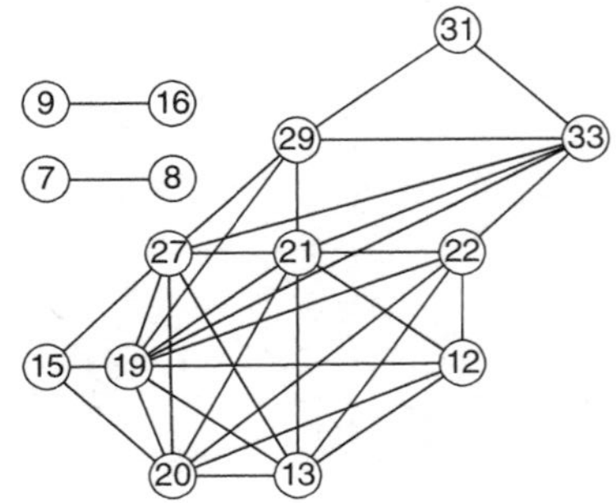

WEB 33: CROCODILE CREEK, NYASA

1 – CROCODILUS NILOTICUS
2 – JUVENILE CICHLIDAE
3 – BARBUS INNOCENS
4 – SERRANOCHROM.ROBUST
5 – NAUCORIS SP.
6 – CLARIAS MELLANDI
7 – FROGS
8 – DYTISCID BEETLES
9 – ANISOPTERID LARVAE
10 – BARBUS PALUDINOSUS
11 – ALESTES IMBERI
12 – MOSQUITO LARVAE
13 – CYCLOPOID COPEPODS
14 – ZYGOPTERID LARVAE
15 – CARIDINA NILOTICA
16 – BARILIUS MICROCEPH.
17 – TILAPIA SHIRANA
18 – GYRAULUS COSTULATUS
19 – CLADOCERA
20 – CHIRONOMID LARVAE
21 – CADDIS LARVAE
22 – BAETID NYMPHS
23 – MICRONECTA
24 – BARBUS SP.
25 – SEGMENTORBIS ANGUST.
26 – LIMNAEA SP.
27 – BOTTOM ALAGAE, DETRITUS
28 – HAPLOCHROMIS SIMILIS
29 – HIGHER PLANTS
30 – CLARIAS MOSSAMBICUS
31 – INDET. FISHES
32 – GERRIDS
33 – TERRESTRIAL INSECTS

FOOD WEB MATRIX

	1	2	3	4	5	6	7	9	10	11	12	13	14	15	16	17	18	19	20	21	22	23	24	25	26	27	28	29
7	1	0	0	0	0	0	0	0	0	0	0	0	0	0	0	0	0	0	0	0	0	0	0	0	0	0	0	0
8	1	0	0	0	0	0	0	0	0	0	0	0	0	0	0	0	0	0	0	0	0	0	0	0	0	0	0	0
9	0	0	0	0	1	1	0	0	0	0	0	0	0	0	0	0	0	0	0	0	0	0	0	0	0	0	0	0
12	0	0	0	1	0	0	0	0	0	0	0	0	0	0	0	0	0	0	0	0	0	0	0	0	0	0	0	0
13	0	0	1	1	0	0	0	1	0	0	0	0	0	0	0	1	0	0	0	0	0	0	0	0	0	0	0	0
15	0	1	0	0	0	0	0	0	0	0	0	0	0	0	0	0	0	0	0	0	0	0	0	0	0	0	0	0
18	0	0	0	0	1	0	0	0	0	0	0	0	0	0	0	0	0	0	0	0	0	0	0	0	0	0	0	0
19	0	1	1	1	0	0	0	0	1	0	0	1	0	0	0	0	0	0	0	0	0	0	0	0	0	0	0	0
20	0	1	1	1	0	0	0	1	0	0	0	0	1	0	0	1	0	0	0	0	0	0	0	0	0	0	0	0
21	0	0	0	1	0	0	0	0	1	0	0	0	0	0	0	0	0	0	0	0	0	0	0	0	0	0	0	0
22	0	0	0	1	0	0	0	0	0	0	0	0	1	0	1	0	0	0	0	0	0	0	0	0	0	0	0	0
27	0	1	1	0	0	0	1	0	1	0	1	1	0	1	0	1	1	1	1	1	1	1	0	1	1	1	0	0
29	0	0	0	0	0	0	0	0	1	0	0	0	0	0	0	0	0	0	0	0	0	0	0	0	0	0	1	0
31	0	0	0	0	0	0	0	0	0	0	0	0	0	0	0	0	0	0	0	0	0	0	0	0	0	0	1	0
33	0	0	0	0	0	0	0	0	1	1	0	0	0	1	0	0	0	0	0	0	0	0	1	0	0	0	1	1

CONSUMER OVERLAP $G_R(C;\lambda)$

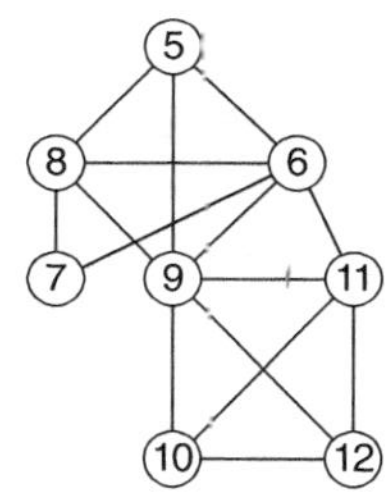

RESOURCE GRAPH $G_C(R;\lambda^{-1})$

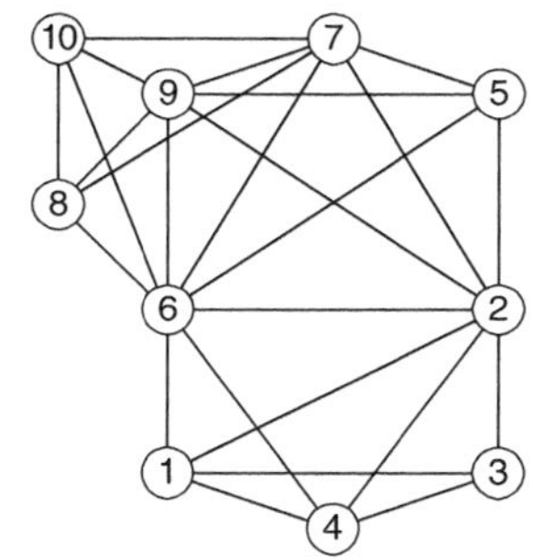

WEB 34: RIVER CLYDACH

1 – LEAF FRAGMENTS
2 – ULOTHRIX, OTHER GREEN ALGAE
3 – DETRITUS
4 – DIATOMS
5 – PROTONEMURA
6 – LEUCTRA, BAETIS, EPHEMERELLA, DIPTERA
7 – PHILOPOTAMUS
8 – ECDYONURUS
9 – HYDROPSYCHE
10 – RHYACOPHILA
11 – PERLA
12 – DINOCRAS

FOOD WEB MATRIX

	5	6	7	8	9	10	11	12
1	1	1	0	1	1	0	0	0
2	0	1	0	0	1	0	1	0
3	0	1	1	1	0	0	0	0
4	0	1	1	1	1	0	0	0
5	0	0	0	0	0	0	1	0
6	0	0	0	0	1	1	1	1
7	0	0	0	0	0	0	1	1
8	0	0	0	0	0	1	0	1
9	0	0	0	0	0	1	1	1
10	0	0	0	0	0	0	0	1

CONSUMER OVERLAP $G_R(C;\lambda)$

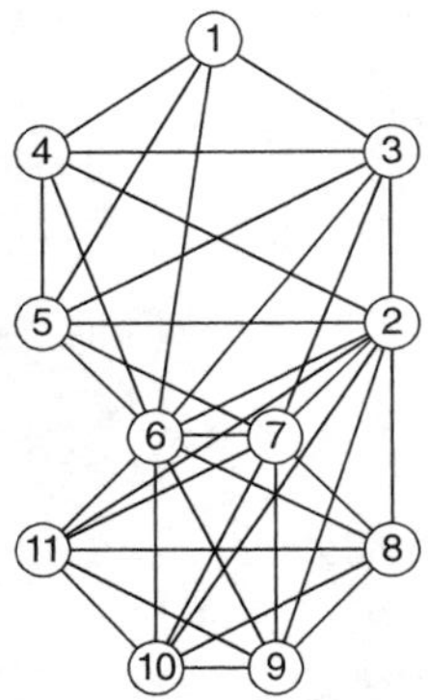

RESOURCE GRAPH $G_C(R;\lambda^{-1})$

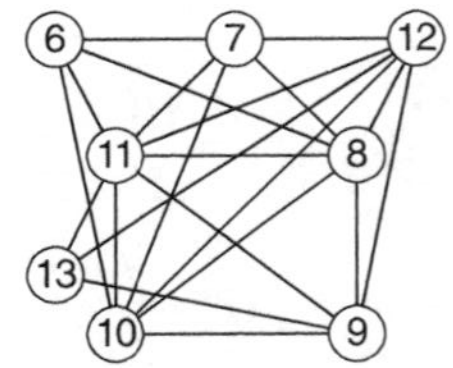

WEB 35: MORGAN'S CREEK

1 – PHAGOCATA
2 – DECAPODA
3 – PLECOPTERA
4 – MEGALOPTERA
5 – PISCES
6 – GAMMARUS
7 – TRICHOPTERA
8 – ASELLUS
9 – EPHEMEROPTERA
10 – TRICHOPTERA
11 – DIPTERA
12 – DETRITUS
13 – DIATOMS

FOOD WEB MATRIX

	1	2	3	4	5	6	7	8	9	10	11
6	1	0	0	1	1	0	0	0	0	0	0
7	0	0	0	1	1	1	0	0	0	0	0
8	1	0	1	1	1	1	0	0	0	0	0
9	0	1	1	0	0	1	1	0	0	0	0
10	0	1	1	1	0	0	0	0	0	0	0
11	0	1	1	0	1	1	1	0	0	0	0
12	0	1	0	0	0	1	1	1	1	1	1
13	0	0	0	0	0	0	1	1	1	1	1

CONSUMER OVERLAP $G_R(C;\lambda)$

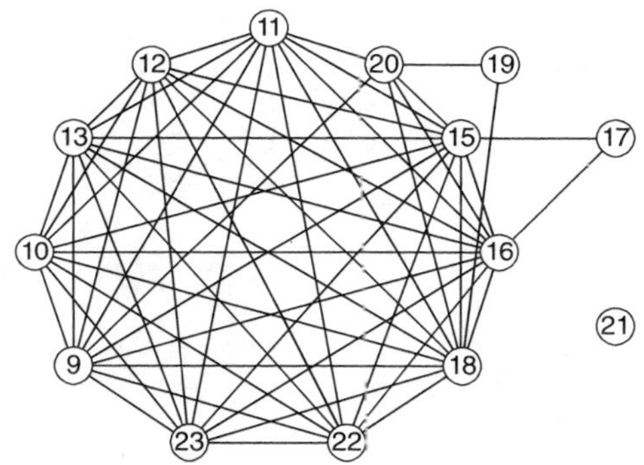

RESOURCE GRAPH $G_C(R;\lambda^{-1})$

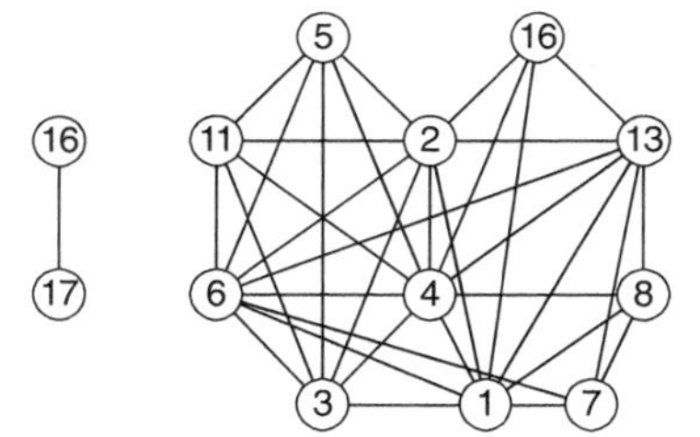

WEB 36: MANGROVE SWAMP

1 – DETRITUS
2 – DIATOMS
3 – CHLAMYDOMONAS
4 – ULOTHRIX
5 – NEMATODES
6 – OSTRACODS
7 – MELAM. PARVULUS
8 – LITTORINA SCABRA
9 – TILAPIA MOSSAMBICA
10 – PODOPHTHAL. VIGIL
11 – COPEPODS
12 – METOP. MESSOR
13 – MACROBRACHIUM
14 – OXYUR. LONCHOTUS
15 – PALAEMONETES
16 – KUHLIA SANDVICENSIS
17 – CHARYBDIS ORIENTALIS
18 – SCYLLA SERRATA
19 – ELEOTRIS SANDWICENSIS
20 – CONGER MARGINATUS
21 – MUGIL CEPHALUS
22 – CHONOPH. GENIVITTATUS

FOOD WEB MATRIX

	9	10	11	12	13	15	16	17	18	19	20	21	22	23
1	1	0	1	0	0	0	0	0	1	0	1	0	0	0
2	1	1	1	1	1	1	1	0	1	0	0	0	1	1
3	0	0	1	1	0	1	0	0	0	0	0	0	0	0
4	0	0	0	1	1	1	1	0	1	0	0	0	0	0
5	0	0	0	0	0	1	0	0	0	0	0	0	0	0
6	0	0	0	0	0	1	1	0	0	0	1	0	0	0
7	0	0	0	0	0	0	0	0	0	0	1	0	0	0
8	0	0	0	0	0	0	0	0	0	0	1	0	0	0
11	0	0	0	0	0	1	1	1	0	0	0	0	0	0
13	0	0	0	0	0	0	0	0	1	1	1	0	0	0
16	0	0	0	0	0	0	0	0	1	1	0	0	0	0
17	0	0	0	0	0	0	0	0	0	0	0	1	0	0
18	0	0	0	0	0	0	0	0	0	0	0	1	0	0

CONSUMER OVERLAP $G_R(C;\lambda)$

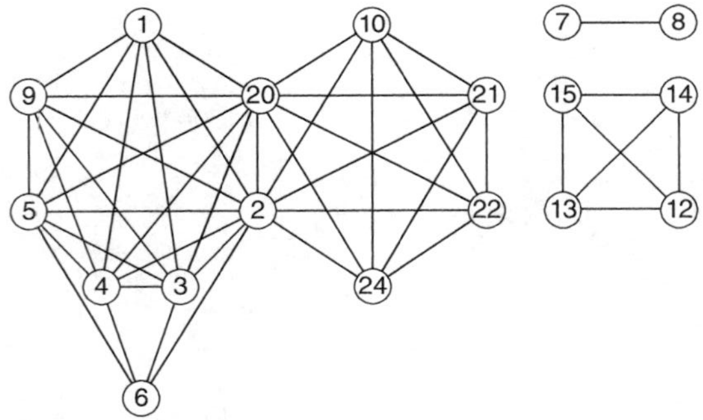

RESOURCE GRAPH $G_C(R;\lambda^{-1})$

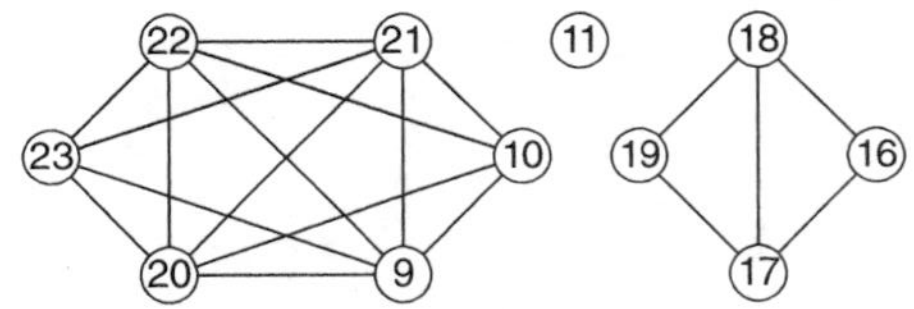

WEB 37: CALIF. SUBLITTORAL

1 – PACIFIC BONITO
2 – VERMILION ROCKFISH
3 – CALIF. SCORPION FISH
4 – CALIFORNIA SEA LION
5 – CABEZON
6 – SAND BASS
7 – PACIFIC ANGEL SHARK
8 – CALIFORNIA HALIBUT
9 – SQUID
10 – OCTOPUS
11 – SMALL FISHES, INVERTS
12 – PACIFIC SAND DAB
13 – PINK SEA PERCH
14 – HORNYHEAD TURBOT
15 – LONGFIN SAND DAB
16 – OPHIUROIDS
17 – POLYCHAETES
18 – BENTHIC CRUSTACEA
19 – HYPOPLANKT. CRUST.
20 – WHITE CROAKER
21 – JACK MACKEREL
22 – NORTHERN ANCHOVY
23 – ZOOPLANKTON
24 – SEA PERCHES

FOOD WEB MATRIX

	1	2	3	4	5	6	7	8	9	10	12	13	14	15	20	21	22	24
9	0	1	1	1	1	1	0	0	0	0	0	0	0	0	0	0	0	0
10	0	0	1	0	1	1	0	0	0	0	0	0	0	0	0	0	0	0
11	0	0	0	0	0	0	1	1	0	0	0	0	0	0	0	0	0	0
16	0	0	0	0	0	0	0	0	0	0	0	0	0	1	0	0	0	0
17	0	0	0	0	0	0	0	0	0	0	0	0	1	1	0	0	0	0
18	0	0	0	0	0	0	0	0	0	0	1	1	1	1	0	0	0	0
19	0	0	0	0	0	0	0	0	0	0	1	0	1	0	0	0	0	0
20	0	1	1	1	0	0	0	0	0	0	0	0	0	0	0	0	0	0
21	1	1	1	1	0	0	0	0	0	0	0	0	0	0	0	0	0	0
22	1	1	1	1	1	0	0	0	1	0	0	0	0	0	1	0	0	0
23	0	1	0	0	0	0	0	0	0	1	0	0	0	0	1	1	1	1

CONSUMER OVERLAP $G_R(C;\lambda)$

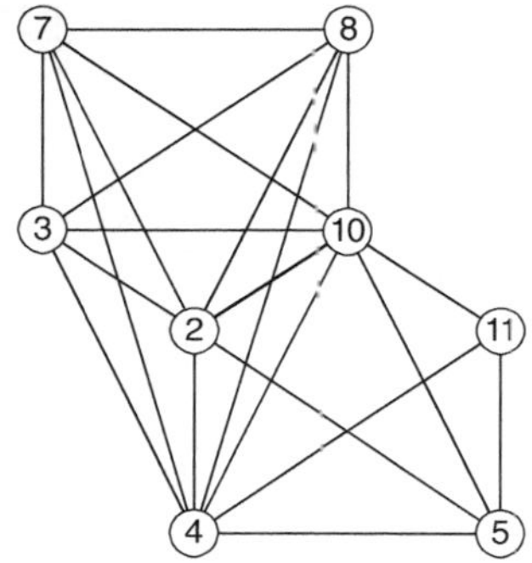

RESOURCE GRAPH $G_C(R;\lambda^{-1})$

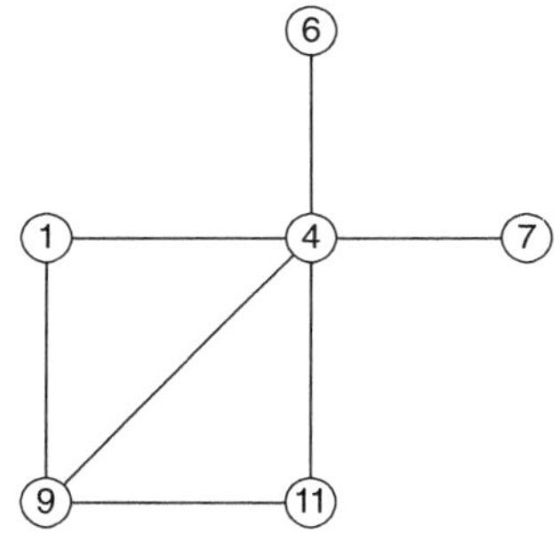

WEB 40: MALAYSIAN RAIN FOREST

1 – CANOPY – LEAVES, FRUITS, FLOWERS
2 – CANOPY ANIMALS – BIRDS, FRUIT BATS, ETC
3 – UPPER AIR ANIMALS – INSECTIVORES, CARNIVORES
4 – INSECTS
5 – LARGE GROUND ANIMALS
6 – TRUNK, FRUIT, FLOWERS
7 – MIDDLE ZONE SCANSORIAL ANIMALS
8 – MIDDLE ZONE FLYING ANIMALS
9 – GROUND – ROOTS, FALLEN FRUIT, LEAVES, ETC
10 – SMALL GROUND ANIMALS
11 – FUNGI

FOOD WEB MATRIX

	2	3	4	5	7	8	10	11
1	1	0	1	1	0	0	0	0
4	1	1	1	0	1	1	1	0
6	0	0	0	0	1	0	0	0
7	0	0	0	0	0	1	0	0
9	0	0	1	1	0	0	1	1
11	0	0	0	0	0	0	1	0

APPENDIX C

Niche Overlap Graphs at Various Thresholds for Cody's (1972) Bird Communities

SONORAN DESERT

$\alpha \geq .65$

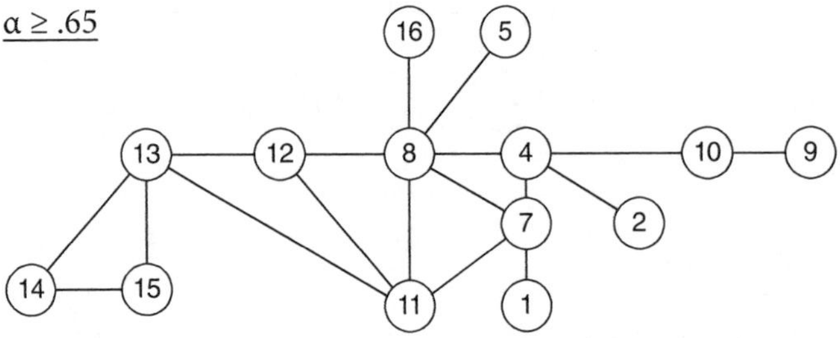

$\alpha \geq .6$

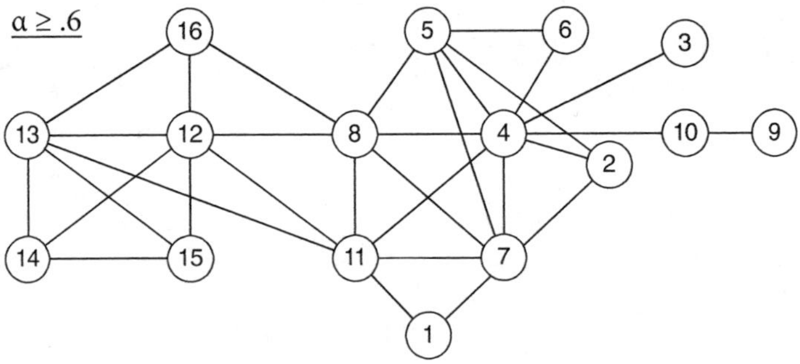

$\alpha \geq .55$

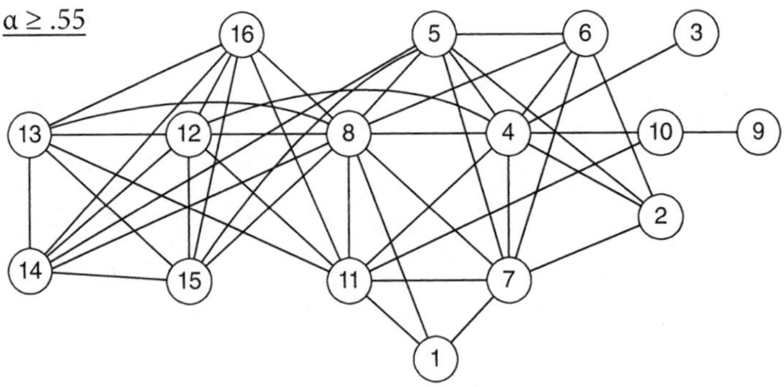

ARIZONA MESQUITE

α ≥ .6

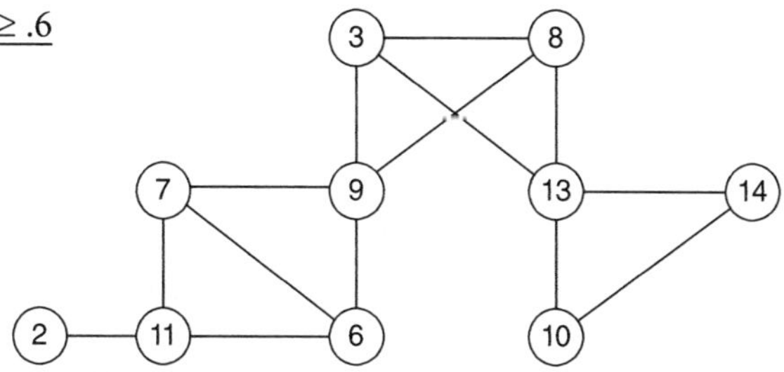

α ≥ .5

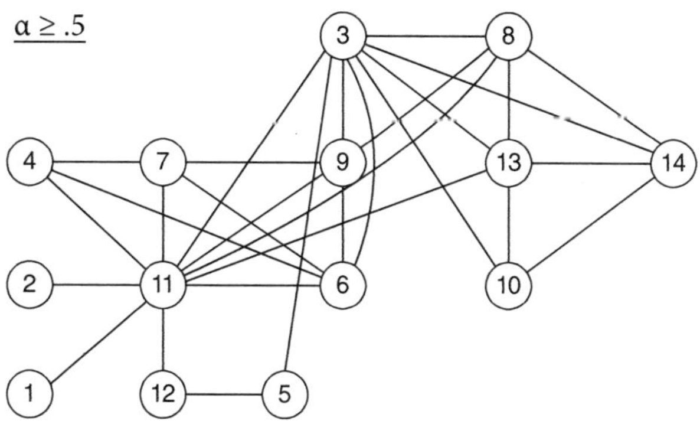

α ≥ .4

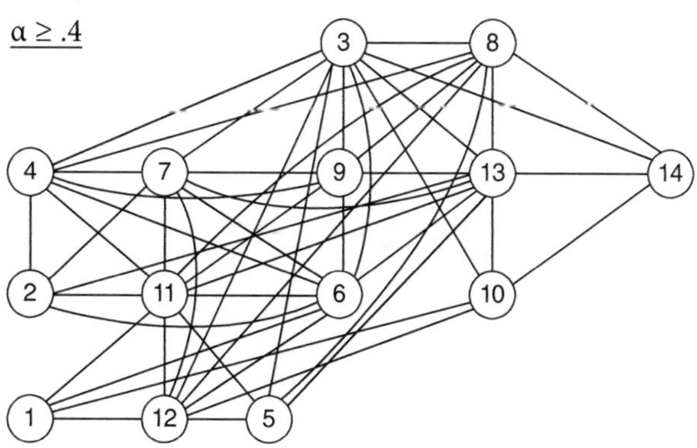

SAVANNAH, CHILE

α > .7

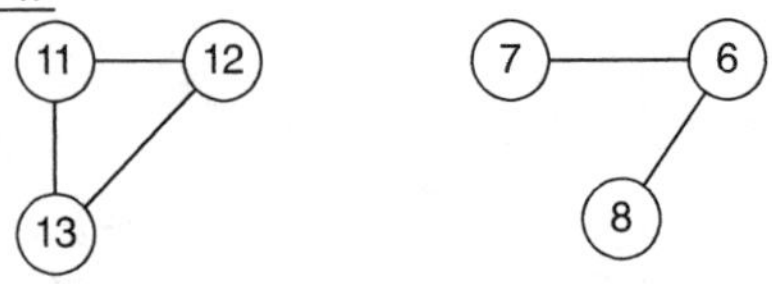

α > .6

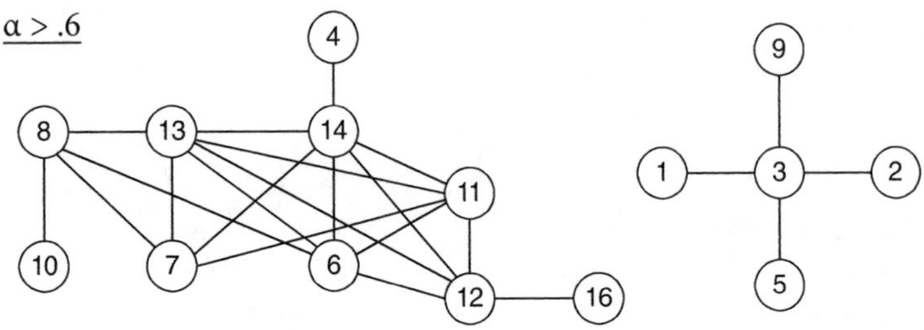

α > .5

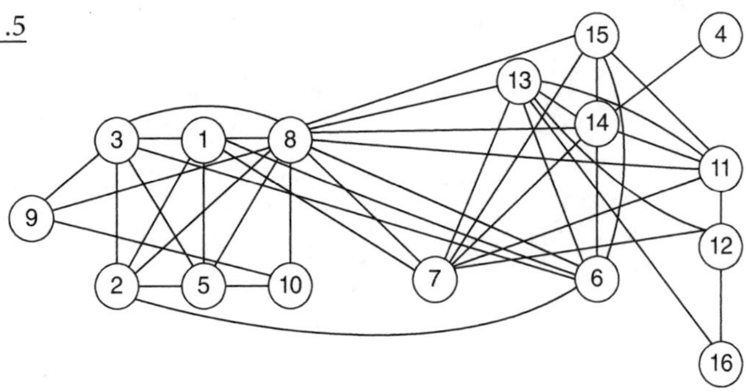

MATORRAL, CHILE

α ≥ .7

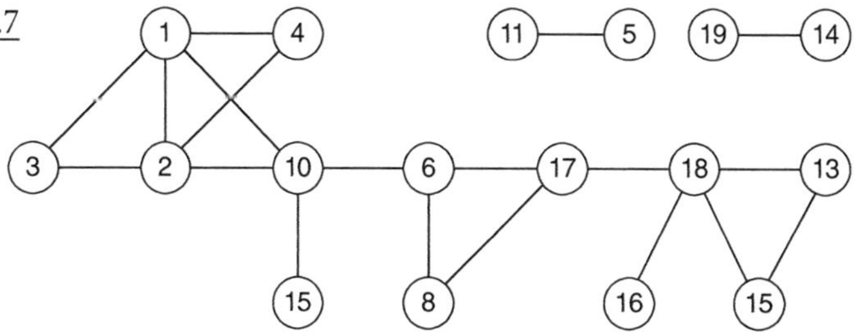

α ≥ .6

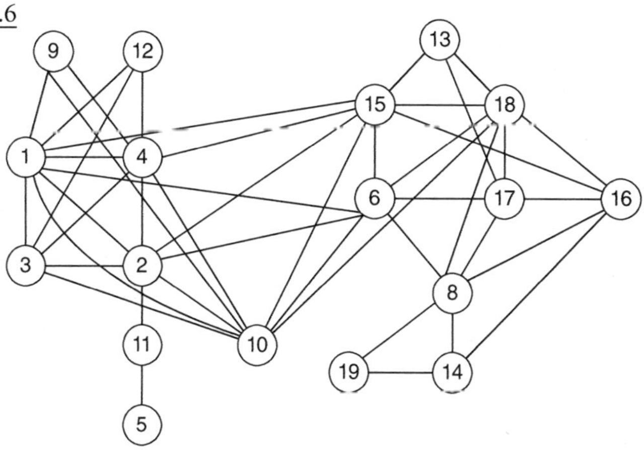

α ≥ .5

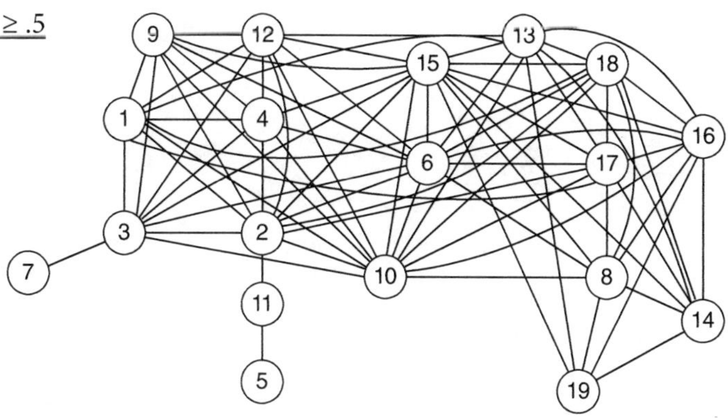

COASTAL SCRUB, CHILE

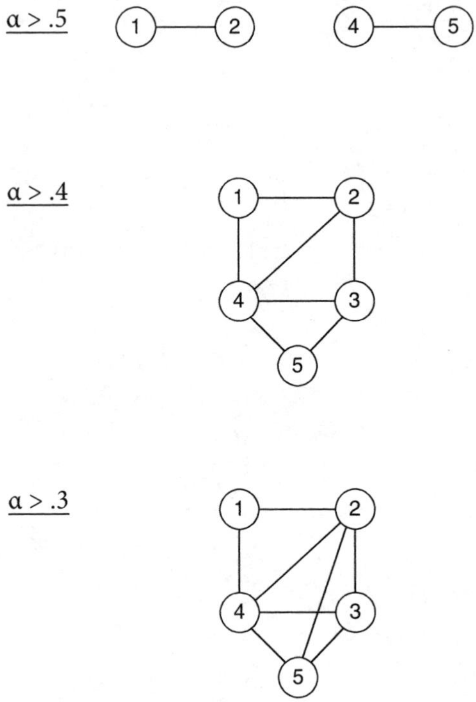

MOHAVE DESERT

$\alpha \geq .6$

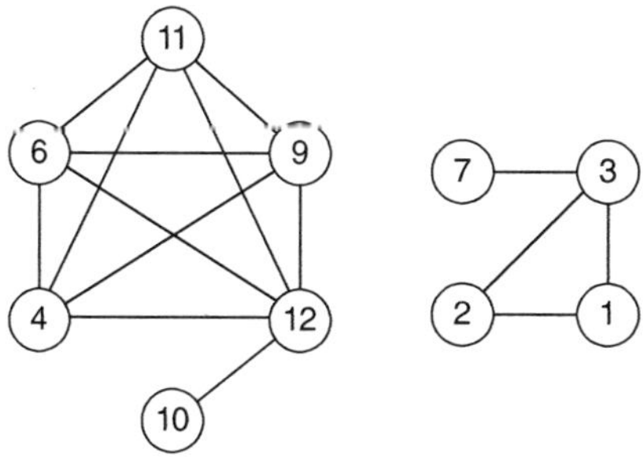

$\alpha \geq .5$

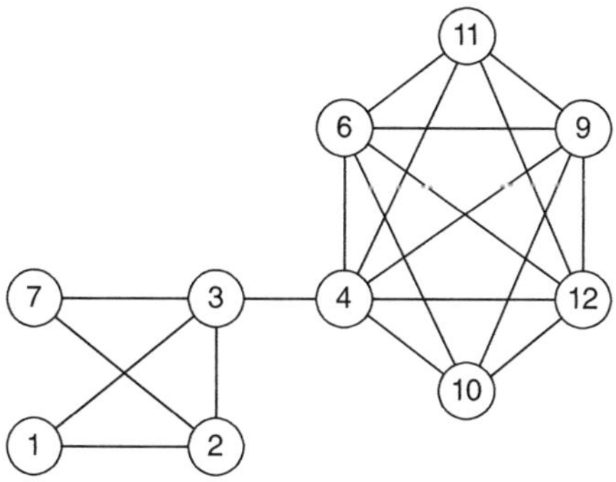

$\alpha \geq .4$

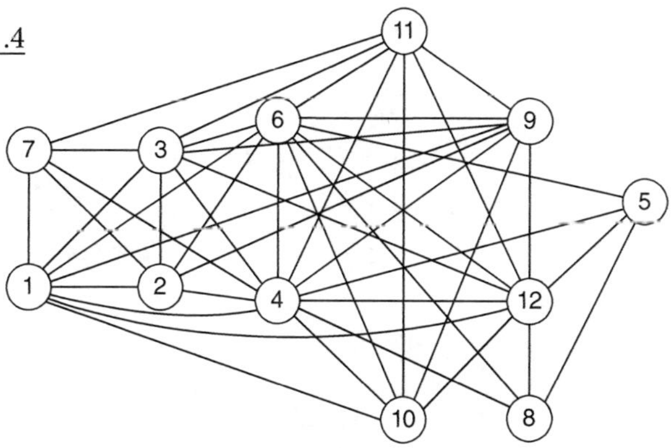

COLORADO SAGEBRUSH

α > .55

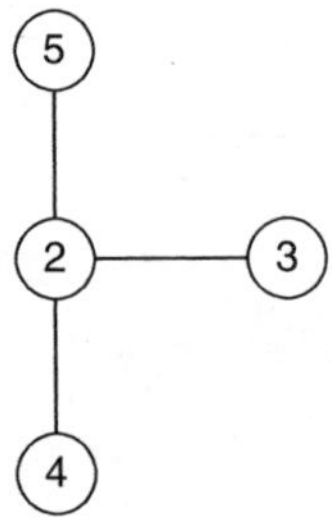

α > .45

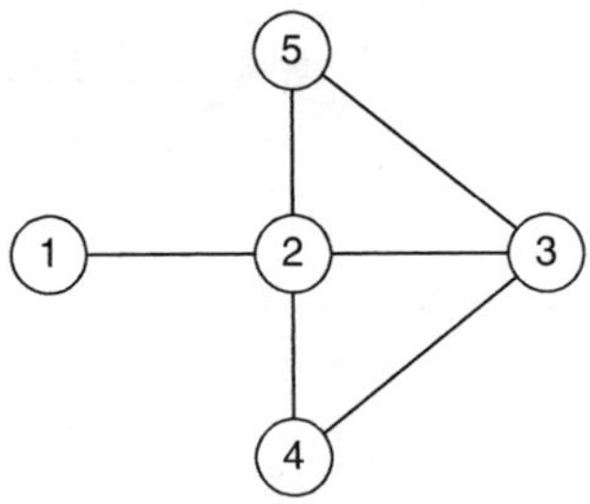

α > .35

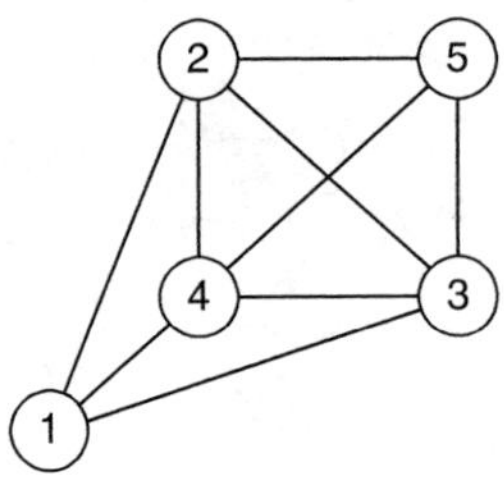

CALIFORNIA CHAPARRAL

$\alpha \geq .7$

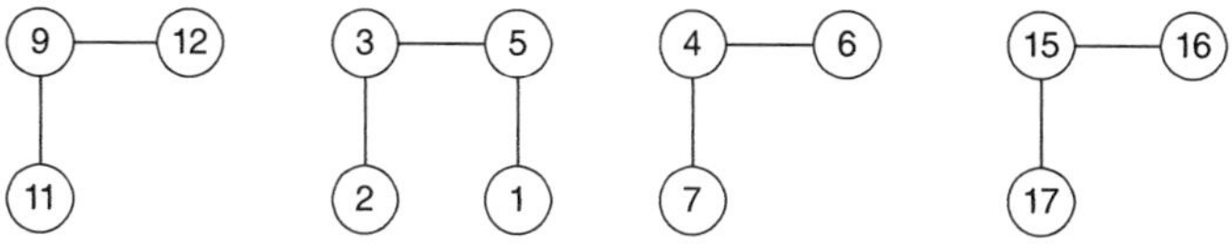

$\alpha \geq .6$

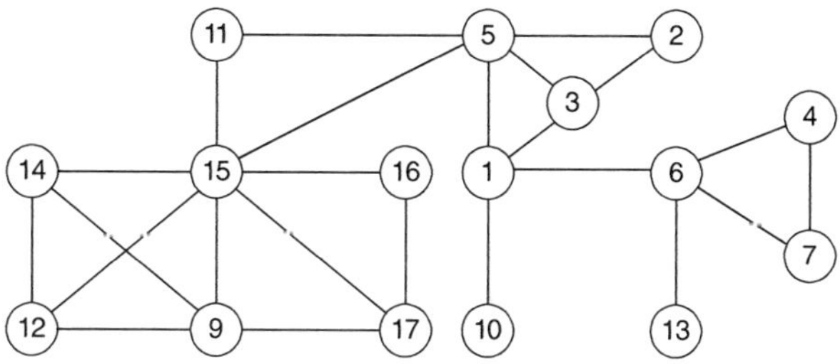

$\alpha \geq .5$

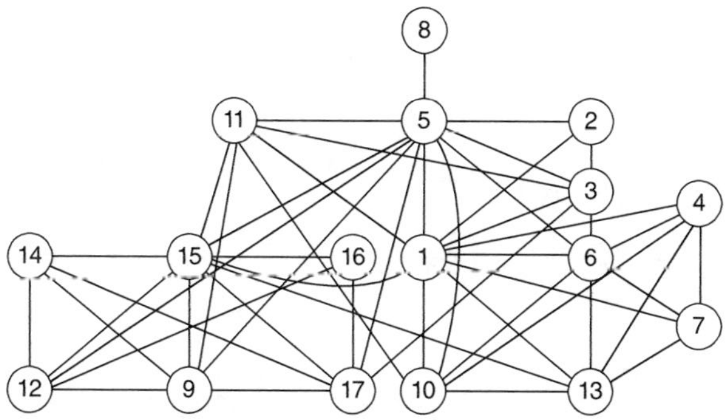

APPENDIX D

Moments of Particle Size Distributions Under Sequential Breakage

By Andrew F. Siegel and George Sugihara

E.1. EXPECTED MEANS AND VARIANCE

Consider a stick of unit length and a breakage distribution F that is symmetric on (0,1). This is stage 1; at stage n the stick will be broken into n pieces. To go from stage n to stage n + 1, first choose a piece at random (uniformly without regard to size, so that each piece has probability 1/n of being chosen) and then break it in two according to a proportion chosen independently from F. Because the piece to be broken is chosen randomly in this way, we lose no generality by requiring F to be symmetric, in order to simplify the mathematical treatment.

If W is an observation from F, define moments $\mu_m = E[\log(W)]^m$ and $\eta = E[\log(W)\log(1 - W)]$ and assume that μ1, μ2, and η are finite. At stage n, let the pieces have sizes $X_{\beta n1n}, \ldots, X_{nn}$ in some random ordering so that these are exchangeable (but not independent) random variables. The logarithms of the sizes, $U_{in} = \log(X_{in})$, are of interest, as are the sample mean

$$\bar{U}_{in} = (\sum_{i=1}^{n} U_{in})/n$$

and the variance

$$S_n^2 = [\sum_{i=1}^{n} (U_{in} - \bar{U}_n)^2] / (n - 1)$$

Notice that each X_{in} is the product of a random number of independent proportions chosen from F, and each U_{in} is the sum of the corresponding logarithms.

Theorem 1. The mean and variance of the logarithm, U_{in}, of the size of a single random piece at stage n are as follows:

$$E(U_{in}) = 2\mu_1 \sum_{k=2}^{n} \frac{1}{k} \tag{1}$$

$$\mathrm{Var}(U_{in}) = 2\mu_2 \sum_{k=2}^{n} \frac{1}{k} - 4_1\mu^2 \sum_{k=2}^{n} \frac{1}{k^2} \tag{2}$$

Proof. By exchangeability, it will suffice to compute $E(U_{in})$. We condition according to whether or not this piece was involved in the most recent breakage, events with probabilities 2/n and (n – 2)/n respectively. Because the conditional distributions of the lengths are $U_{1,n-1}$ log(W) and $U_{1,n-1}$ respectively, where W has distribution F and is independent of $U_{1,n-1}$, we obtain the recurrence

$$E(U_{in}) = E(U_{1,n-1}) + \frac{2}{n}\mu_1 \tag{3}$$

whose solution with initial condition $E(U_{1,1}) = 0$ is (1). For the variance, $\mathrm{Var}(U_{in}) = E(U_{1n}^2) - [E(U_{1n})]^2$, condition as before for each term of this difference, to establish

$$E(U_{1n}^2) = E(U_{1,n-1}^2) + \frac{4}{n}\mu_1 E(U_{1,n-1}) + \frac{2}{n}\mu_2 \tag{4}$$

and

$$E[(U_{1n})]^2 = [E(U_{1,n-1})]^2 + \frac{4}{n}\mu_1 E(U_{1,n-1}) + \frac{4^2}{n^2}\mu_1 \tag{5}$$

Subtracting (5) from (4) and simplifying, we obtain

$$\mathrm{Var}(U_{1n}) = \mathrm{Var}(U_{1,n-1}) + \frac{2}{n}\mu_2 - \frac{4\mu_1^2}{n^2} \tag{6}$$

whose solution with initial condition $\mathrm{Var}(U_{11}) = 0$ is (2).

In a practical setting, care must be taken regarding the variance term. Due to the dependence among $U_{1n}, \ldots, U_{nn}$, the sample variance S_n^2 should not be compared to (2), which is the expected sample variance for an *independent* sample with the same marginals. Instead, the following quantities should be used.

Theorem 2. The expected sample mean and variance at stage n are

$$E(\overline{U}_n)=2\mu_1 \sum_{k=2}^{n} \frac{1}{k} \tag{7}$$

$$E(S_n^2)=\{2(1+\frac{2}{n-1}) \sum_{k=2}^{n} \frac{1}{k} - 2\}\mu_2 - \eta \tag{8}$$

Proof. Equation (7) follows by linearity from (1). For (8), expand and use exchangeability to obtain

$$E(S_n^2)=E(U_{1n}^2 - U_{1n}U_{2n}). \tag{9}$$

Next, condition according to the four possibilities of involvement of U_{1n} and U_{2n} in the most recent breakage (neither, U_{1n} only, U_{2n} only, or both). For example, with probability 2/(n(n – 1)), they were both involved in the most recent breakage, and $U_{1n}^2 - U_{1n}U_{2n}$ has the conditional distribution

$$[U_{1,n-1} + \log(W)]^2 - \{[U_{1,n-1} + \log(W)][U_{1,n-1} + \log(1-W)]\} \tag{10}$$

where W is a random variable with distribution F and is independent of (U_{1n}, U_{2n}). Combining this with the results from the other three possibilities, then simplifying, we find with some effort that

$$E(S_n^2) = \frac{(n-2)(n+1)}{n(n-1)} E(S_{n-1}^2) + \frac{2}{n}\mu_2 - \frac{2\eta}{n(n-1)} \tag{11}$$

With some patience, it can be shown by induction that (8) is the solution to the recurrence (11) with initial condition $E(S_2^2) = \mu_2 - \eta$. The means (1) and (7) are identical because the expectation operator is linear even under dependence. However, from (2) and (8) we see that the lack of independence has modified the variance. These differences can be studied in detail by examining an asymptotic expansion of each expression.

Theorem 3.

$$E(U_{in})=E(U_n)=\mu_1\{2\log(n)-0.8456+\frac{1}{n}-\frac{1}{6n^2}+\frac{1}{60n^4}+0\frac{1}{n^6}\}$$

$$Var(U_{in})=2\mu_2\log(n)-(2.5797\mu_2^2+0.8456\mu_2) \tag{13}$$

$$\frac{4\mu_1^2+\mu_2}{n}+\frac{12\mu_1^2+\mu_2}{6n^2}+\frac{2\mu_1^2}{3n^3}+\frac{\mu_2}{60n^4}-\frac{2\mu_1^2}{15n^5}\frac{0(1)}{n^6}$$

$$E(S_n^2)= 2\mu_2 \log(n)-(\eta+2.8456\mu_2)+4\mu_2 \frac{\log(n)}{n-1}-\frac{0.6911\mu_2}{n} \qquad (14)$$

$$+\frac{0.1422\mu_2}{n^2}-\frac{0.02447\mu_2}{n^3}-\frac{0.007804\mu_2}{n^4}+\frac{0.008863\mu_2}{n^5}\frac{0(1)}{n^6}$$

Proof. These follow from two standard asymptotic expansions. Equation (12) depends on the expansion of the harmonic series (e.g., Knuth, 1973, Vol. 1, p. 74):

$$\sum_{k=1}^{n}\frac{1}{k}=\log(n)+\gamma+\frac{1}{2n}-\frac{1}{12n^2}+\frac{1}{120n^4}+0\frac{1}{n^6} \qquad (15)$$

where $\gamma \cong 0.5772156649$ is Euler's constant. Equation (13) also uses (e.g., p. 61 of Hansen, 1975)

$$\sum_{k=1}^{n}\frac{1}{k^2}=\frac{\pi^2}{6}-\frac{1}{n}-\frac{1}{2n^2}-\frac{1}{6n^3}+\frac{1}{30n^5}+0\frac{1}{n^7} \qquad (16)$$

For (14), use (15) in (8) and multiply the nonlogarithmic part by the expansion of 2/(n – 1) as a power series in 1/n. Combine terms to find

$$E(S_n^2)=\{2\log(n)+(2\gamma-4)+\frac{4\log(n)}{n-1}+\frac{4\gamma-3}{n} \qquad (17)$$

$$\frac{24\gamma-13}{6n^2}+\frac{24\gamma-14}{6n^3}+\frac{240\gamma-139}{60n^4}+\frac{240\gamma-138\}}{60n^5}\mu_2-\eta+0\frac{1}{n^6}$$

which evaluates to (14).

Although the leading terms in the variances (13) and (14) are identical, the differences in their second terms cannot be neglected even for moderately large n due to the slowly increasing behavior of log(n). For example, when F places mass ½ at [[frac14]] and at [[frac34]], even with n as large as 50, we have $\mathrm{Var}(U_{in}) = 5.3$ while $E(S_n^2)=4.9$.

E.2. AN IDENTITY FOR RANDOM BINARY TREES

Consider the class of random binary trees with n endpoints generated recursively by bifurcating an endpoint chosen uniformly at random from a tree with n – 1 endpoints. These trees are responsible for part of the randomness for the sequential breakage model (the other component can be thought of as entering through the distribution F). These trees are related to random binary search trees used in computer science (Knuth, 1973, vol. 3, p. 423–471). For a tree with n endpoints, let N_{1n} and N_{2n} denote the distances (in numbers of edges) from each of two randomly chosen endpoints to their nearest common ancestor, as illustrated in the Figure 7.8. Although moments involving N_{1n} and N_{2n} generally increase with n, there is an expression for which this dependence cancels out. This leads to the following key theorem used in chapter 7 (7.2).

Theorem 4. Regardless of the value of n,

$$E(N_{1n} + N_{1n}N_{2n} - N_{1n}^2) = 1 \quad (18)$$

Proof. Proceed by induction on n, conditioning on the four events describing which of N_{1n} and N_{2n} were involved in the most recent bifurcation. This is similar to the proof of Theorem 2.

REFERENCES

Atkin, R.H. 1972. Multi-dimensional structure in the game of chess. Int. J. Man-Machine St., 4:139–167, 341–362.

Atkin, R.H. 1974. An algebra for patterns and complexes. Int. J. Man-Machine St., 6:285–307.

Auerbach, M.S. 1982. Stability, probability and the topology of food webs. (Manuscript submitted).

Booth, K.S. and G.S. Leuker. 1976. Testing for the consecutive ones property, interval graphs, and graph planarity using p-q-tree algorithms. J. Comput. Syst. Sci., 13:335–379.

Briand, F. 1982. Environmental control of food web structure. Ecology (in press).

Buneman, P. 1975. A characterization of rigid circuit graphs. Discrete Math., 4:235.

Casti, J. 1979. *Connectivity, Complexity, and Catastrophe in Large Scale Systems.* Wiley Interscience, ISASA.

Clements, F.E. 1916. *Plant Succession, an Analysis of the Development of Vegetation.* Carnegie Institution of Washington Publication, 242:1–512.

Cody, M. 1974. *Competition and the Structure of Bird Communities.* Princeton University Press.

Cohen, J.E. 1978. *Food Webs and Niche Space.* Princeton Univ. Press, Princeton.

Colwell, R.K. and E.R. Fuentes. 1975. Experimental studies of the niche. Annu. Rev. Ecol. Syst., 6:281–310.

Curtis, J.T. 1954. *The Vegetation of Wisconsin.* Univ. of Wisconsin Press, Madison.

Darlington, P.J. 1957. *Zoogeography.* New York: John Wiley and Sons.

Day, J.H. 1967. The biology of the Knysna Estuary, South Africa. In *Estuaries,* G.H. Lauff, ed. (AAAS publication 83) Washington, 397–407.

De Angelis, D.H. 1975. Stability and connectance in food web models. Ecology, 56:238–243.

Diamond, J.M. 1975. Assembly of species communities. in *Ecology and Evolution of Communities*, M.L. Cody, J.M. Diamond, eds., pp 342–444.

Dirac G.A. 1961. On rigid circuit graphs, Abh. Math. Sem. Univ. Hamburg, 25:71–76.

Fulkerson, D.R. and O.A. Gross. 1965. Incidence matrices and interval graphs. Pacific J. Math., 15:835–855.

Gallopin, G.C. 1972. Structural properties of food webs. Syst. Anal. β Simul. in Ecol. II (ed. by B.C. Patten), Academic Press, New York, 241–282.

Gardner, M.R. and W.R. Ashby. 1970. Connectance of large dynamical (cybernetic) systems: critical valves for stability. Nature, 228:784.

Gatz, A.J., Jr. 1979. Community organization in fishes as indicated by morphological features. Ecology, 60:711–718.

Giblin, P.J. 1977. *Graphs Surfaces and Homology*. Chapman and Hall, London.

Gilpin, M.E. and F.J. Ayala. 1973. Proc. Nat. Acad. Sci., 70:3590.

Gilpin, M.E. 1975. Stability of feasible predator-prey systems. Nature, 254:137–138.

Goh, B.S. 1978. Robust stability concepts for ecosystem models. *Theoretical Systems Ecology*, ed. E. Halfon, Academic Press, New York.

Hairston, N.G. and G.W. Byers. 1954. The soil arthropods of a field in southern Michigan: a study in community ecology. Contrib. Lab. Vertebr. Biol. Univ. Mich., 64:1–37.

Happ, H.H. 1971. *Diakoptics and Networks*. Academic Press, New York.

Harary, F. 1960. A graph theoretic method for the complete reduction of a matrix with a view toward finding its eigenvalues. J. Math. and Phys., 38:104–111.

Harary, F. 1971. Sparse matrices and graph theory, in *Large Sparse Sets of Linear Equations*. ed. Reid, Academic Press, London, 139–149.

Harary, F. 1972. *Graph Theory*. Addison-Wesley.

Heatwole, H. and R. Lwvins. 1972. Trophic structure stability and faunal change during recolonization. Ecology, 53:531–534.

Hilton, P.J. and S. Wylie. 1960. Homology Theory. Cambridge University Press, London.

Holmes, R.T., Bonney, R.E. and S.W. Pacala. 1979. Guild structure of the Hubbard Brook bird community: a multivariate approach. Ecology, 60:512–520.

Horn, H.S. 1971. *The Adaptive Geometry of Trees*. Princeton Univ. Press, Princeton.

Hurlbert, S.H. 1981. A gentle depilation of the niche: Dicean resource sets in resource hyperspace. Evol. Theory, 5:177–184.

Hutchinson, G.E. 1958. Concluding remarks. Cold Spring Harbor Sympos. Quant. Biol., 22:415–422.

Hutchinson, G.E. 1967. *A Treatise on Limnology*. Vol. II. Wiley, New York, 115 pp.

Kolmogoroff, A.N. 1941. Uber das logarithmisch normale Verteilungsgestz der Dimensionen der Teilchen bei Zerstuckelung. C.R. (Doklady) Acad. Sci. URSS, 31:99–101.

Levins, R. 1974. The qualitative analysis of partially specified systems. Ann. N.Y. Acad. Sci., 231:123–138.

Lindemann, R.L. 1942. The trophic-dynamic aspect of ecology. Ecology, 23:399–418.

MacArthur, R.H. 1957. On the relative abundance of bird species. Proc. Nat. Acad. Sci. U.S., 43:293–295.

MacArthur, R.H. 1969. Species packing and what competition minimizes. P.N.A.S., 64:1369–1371.

MacArthur, R.H. 1970. Species packing and competitive equilibrium for many species. Theor. Pop. Biol., 1:1–11.

MacArthur, R.H. and E.O. Wilson. 1967. *The Theory of Island Biogeography*. Princeton University Press, Princeton, N.J.

MacDonald, N. 1979. Simple aspects of foodweb complexity. J. Theor. Biol., 80:577–588.

MacLane, S. 1963. *Homology*. Springer-Verlag, New York.

Margalef, D.R. 1958. Gen. Syst., 3:36–71.

Margalef, R. 1961. On certain unifying principles in ecology. Am. Natur., 97:357–374.

May, R.M. 1974. How many species: some mathematical aspects of the dynamics of populations. Lectures on Mathematics in the Life Sciences 6:65–98.

May, R.M. 1979. The structure and dynamics of ecological communities. In Population Dynamics. (eds. R.M. Anderson and B.D. Turner.)

May, R.M. 1972. Will a large complex system be stable? Nature, 238:413–414.

May, R.M. 1973. *Stability and Complexity in Model Ecosystems*. Princeton University Press, Princeton.

May, R.M. 1975. Some notes on estimating the competition matrix, α, Ecology, 56:737–741.

McNaughton, S.J. 1978. Stability and diversity of ecological systems. Nature, 274:251–253.

Motomura, I. 1932. A statistical treatment of associations. Jpn. J. Zool., 44:379–383.

Nunney, L. 1980. The stability of complex model ecosystems. Am. Natur., 115:639–649.

Parter, S. 1960. The eigenvalues and eigenvectors of a class of matrices. J. Soc. Indust. Appl. Math., 8:376–388.

Patrick, R. 1968. The structure of diatom communities in similar ecological conditions. Am. Nat., 102:173–183.

Patrick, R., Hohn, M. and J. Wallace. 1954. A new method of determining the pattern of the diatom flora. Not. Nat., 259.

Pielou, E.C. 1975. Ecological diversity. Wiley, New York.

Pimm, S.L. and J. H. Lawton. 1980. Are food webs divided into compartments? J. Anim. Ecol., 49:879–898.

Pimm, S.L. 1979. Structure of food webs. Theor. Pop. Biol., 16:144–158.

Post, W.M. and S.L. Pimm. 1982. Community assembly and food web stability. (Manuscript submitted).

Preston, F.W. 1962. The canonical distribution of commonness and rarity. Ecology, 43:185–215, 410–432.

Qazim, S.Z. 1970. In *Marine Food Chains*, (J.H. Steele, ed.). Oliver Boyd, U.K.

Rejmanek, M. and P. Stary. 1979. Connectance in real biotic communities and critical valves for stability in model communities. Nature, 280:311–313.

Ricklefs, R.E. and J. Travis. 1980. A morphological approach to the study of avian community organization, The Auk, 97:321–338.

Roberts, A.P. 1974. Stability of a feasible random ecosystem. Nature, 251:607–608.

Roberts, F.S. 1976. *Discrete Mathematical Models, with Applications to Social, Biological and Environmental Problems*. Prentice-Hall, N.J.

Rose, D. 1970. Triangulated graphs and the elimination process, J. Appl. Math., 32:597.

Rose, D.J., Tarjan, R.E. and G.S. Leuker. 1976. Algorithmic aspects of vertex elimination on graphs, SIAM J. Comput., 5:266–283.

Siegel, A. and G. Sugihara. 1982. Moments of particle size distributions under sequential breakage. J. Appl. Statistics (in press).

Siljak, D.D. 1976. Competitive economic systems: stability, decomposition, and aggregation. IEEE Transactions on Autom. Contrl. 21:149–160.

Siljak, D.D. 1975. When is a complex ecosystem stable? Math. Biosci., 25:25–50.

Stewart, D.B. 1965. Partioning and tearing systems of equations. J. SIAM Numer. Anal., 2:345–365.

Strobeck, C. 1973. N species competion. Ecology, 54:650–654.

Sugihara, G. 1980. Minimal community structure: an explanation of species abundance patterns. Am. Natur., 116:770–787.

Sugihara, G. 1981. A reply to Connor and McCoy. Am. Natur. 117:790–793.

Sugihara, G. 1982. Commentary on diversity as a concept. Jour. Amer. Stat. Assoc., (in press).

Summerhayes, V.S. and C.S. Elton. 1923. Contributions to the ecology of Spitzbergen and Bear Island. J. Ecol., 11:214–286.

Tarjan, R.E. 1978. Private manuscript, Bell Labs Technical Bulletin.

Tregonning, K. and A. Roberts. 1978. Ecosystem-like behavior of a random-interaction model. I. Bull. Math. Biol., 40:513–524.

Tregonning, K. and A. Roberts, 1979. Complex systems which evolve towards homeostasis. Nature, 281:563–564.

Turelli, M. 1926. A reexamination of stability in randomly varying versus deterministic environments with comments on the stochastic theory of limiting similarity. Theor. Pop. Biol., 13:244–262.

Varley, G.C. 1920. *Animal Populations in Relation to their Food Resources*, A. Watson, ed. Blackwell, Oxford, 389.

Wallace, A.H. 1970. *Algebraic Topology, Homology and Cohomology.* Benjamin Inc., New York.

Walsh, G.E. 1967. An ecological study of a Hawaiian mangrove swamp. In *Estuaries*, G.H. Lauff, ed. (AAAS publ. No. 83.) Washington, 420–431.

Whittaker, R.H. 1972. Evolution and measurement of species diversity. Taxon, 21:213–251.

Whittaker, R.H. 1977. Evolution of species diversity in land communities. Evol. Biol., 10:1–66.

Woodwell, G.M. 1967. Sci. Amer., 216:24.

Yodzis, P. 1980. The connectance of real ecosystems. Nature, 284:544–545.

Yodzis, P. 1981. The structure of assembled communities. J. Theor. Biol., 92:103–117.